Coral Reefs: A Very Short Introduction

VERY SHORT INTRODUCTIONS are for anyone wanting a stimulating and accessible way into a new subject. They are written by experts, and have been translated into more than 40 different languages.

The Series began in 1995, and now covers a wide variety of topics in every discipline. The VSI library now contains over 350 volumes—a Very Short Introduction to everything from Psychology and Philosophy of Science to American History and Relativity—and continues to grow in every subject area.

Very Short Introductions available now:

ACCOUNTING Christopher Nobes
ADVERTISING Winston Fletcher
AFRICAN HISTORY John Parker and Richard Rathbone
AFRICAN RELIGIONS Jacob K. Olupona
AGNOSTICISM Robin Le Poidevin
AMERICAN HISTORY Paul S. Boyer
AMERICAN IMMIGRATION David A. Gerber
AMERICAN LEGAL HISTORY G. Edward White
AMERICAN POLITICAL PARTIES AND ELECTIONS L. Sandy Maisel
AMERICAN POLITICS Richard M. Valelly
THE AMERICAN PRESIDENCY Charles O. Jones
ANAESTHESIA Aidan O'Donnell
ANARCHISM Colin Ward
ANCIENT EGYPT Ian Shaw
ANCIENT GREECE Paul Cartledge
THE ANCIENT NEAR EAST Amanda H. Podany
ANCIENT PHILOSOPHY Julia Annas
ANCIENT WARFARE Harry Sidebottom
ANGELS David Albert Jones
ANGLICANISM Mark Chapman
THE ANGLO-SAXON AGE John Blair
THE ANIMAL KINGDOM Peter Holland
ANIMAL RIGHTS David DeGrazia
THE ANTARCTIC Klaus Dodds
ANTISEMITISM Steven Beller
ANXIETY Daniel Freeman and Jason Freeman
THE APOCRYPHAL GOSPELS Paul Foster
ARCHAEOLOGY Paul Bahn
ARCHITECTURE Andrew Ballantyne
ARISTOCRACY William Doyle
ARISTOTLE Jonathan Barnes
ART HISTORY Dana Arnold
ART THEORY Cynthia Freeland
ASTROBIOLOGY David C. Catling
ATHEISM Julian Baggini
AUGUSTINE Henry Chadwick
AUSTRALIA Kenneth Morgan
AUTISM Uta Frith
THE AVANT GARDE David Cottington
THE AZTECS Davíd Carrasco
BACTERIA Sebastian G. B. Amyes
BARTHES Jonathan Culler
THE BEATS David Sterritt
BEAUTY Roger Scruton
BESTSELLERS John Sutherland
THE BIBLE John Riches
BIBLICAL ARCHAEOLOGY Eric H. Cline
BIOGRAPHY Hermione Lee
THE BLUES Elijah Wald
THE BOOK OF MORMON Terryl Givens
BORDERS Alexander C. Diener and Joshua Hagen
THE BRAIN Michael O'Shea
THE BRITISH CONSTITUTION Martin Loughlin
THE BRITISH EMPIRE Ashley Jackson
BRITISH POLITICS Anthony Wright
BUDDHA Michael Carrithers

BUDDHISM Damien Keown
BUDDHIST ETHICS Damien Keown
CANCER Nicholas James
CAPITALISM James Fulcher
CATHOLICISM Gerald O'Collins
CAUSATION Stephen Mumford and Rani Lill Anjum
THE CELL Terence Allen and Graham Cowling
THE CELTS Barry Cunliffe
CHAOS Leonard Smith
CHILDREN'S LITERATURE Kimberley Reynolds
CHINESE LITERATURE Sabina Knight
CHOICE THEORY Michael Allingham
CHRISTIAN ART Beth Williamson
CHRISTIAN ETHICS D. Stephen Long
CHRISTIANITY Linda Woodhead
CITIZENSHIP Richard Bellamy
CIVIL ENGINEERING David Muir Wood
CLASSICAL LITERATURE William Allan
CLASSICAL MYTHOLOGY Helen Morales
CLASSICS Mary Beard and John Henderson
CLAUSEWITZ Michael Howard
CLIMATE Mark Maslin
THE COLD WAR Robert McMahon
COLONIAL AMERICA Alan Taylor
COLONIAL LATIN AMERICAN LITERATURE Rolena Adorno
COMEDY Matthew Bevis
COMMUNISM Leslie Holmes
THE COMPUTER Darrel Ince
THE CONQUISTADORS Matthew Restall and Felipe Fernández-Armesto
CONSCIENCE Paul Strohm
CONSCIOUSNESS Susan Blackmore
CONTEMPORARY ART Julian Stallabrass
CONTEMPORARY FICTION Robert Eaglestone
CONTINENTAL PHILOSOPHY Simon Critchley
CORAL REEFS Charles Sheppard
COSMOLOGY Peter Coles
CRITICAL THEORY Stephen Eric Bronner
THE CRUSADES Christopher Tyerman
CRYPTOGRAPHY Fred Piper and Sean Murphy
THE CULTURAL REVOLUTION Richard Curt Kraus
DADA AND SURREALISM David Hopkins
DARWIN Jonathan Howard
THE DEAD SEA SCROLLS Timothy Lim
DEMOCRACY Bernard Crick
DERRIDA Simon Glendinning
DESCARTES Tom Sorell
DESERTS Nick Middleton
DESIGN John Heskett
DEVELOPMENTAL BIOLOGY Lewis Wolpert
THE DEVIL Darren Oldridge
DIASPORA Kevin Kenny
DICTIONARIES Lynda Mugglestone
DINOSAURS David Norman
DIPLOMACY Joseph M. Siracusa
DOCUMENTARY FILM Patricia Aufderheide
DREAMING J. Allan Hobson
DRUGS Leslie Iversen
DRUIDS Barry Cunliffe
EARLY MUSIC Thomas Forrest Kelly
THE EARTH Martin Redfern
ECONOMICS Partha Dasgupta
EDUCATION Gary Thomas
EGYPTIAN MYTH Geraldine Pinch
EIGHTEENTH-CENTURY BRITAIN Paul Langford
THE ELEMENTS Philip Ball
EMOTION Dylan Evans
EMPIRE Stephen Howe
ENGELS Terrell Carver
ENGINEERING David Blockley
ENGLISH LITERATURE Jonathan Bate
ENTREPRENEURSHIP Paul Westhead and Mike Wright
ENVIRONMENTAL ECONOMICS Stephen Smith
EPIDEMIOLOGY Rodolfo Saracci
ETHICS Simon Blackburn
ETHNOMUSICOLOGY Timothy Rice
THE ETRUSCANS Christopher Smith
THE EUROPEAN UNION John Pinder and Simon Usherwood
EVOLUTION Brian and Deborah Charlesworth
EXISTENTIALISM Thomas Flynn
THE EYE Michael Land

FAMILY LAW Jonathan Herring
FASCISM Kevin Passmore
FASHION Rebecca Arnold
FEMINISM Margaret Walters
FILM Michael Wood
FILM MUSIC Kathryn Kalinak
THE FIRST WORLD WAR Michael Howard
FOLK MUSIC Mark Slobin
FOOD John Krebs
FORENSIC PSYCHOLOGY David Canter
FORENSIC SCIENCE Jim Fraser
FOSSILS Keith Thomson
FOUCAULT Gary Gutting
FRACTALS Kenneth Falconer
FREE SPEECH Nigel Warburton
FREE WILL Thomas Pink
FRENCH LITERATURE John D. Lyons
THE FRENCH REVOLUTION William Doyle
FREUD Anthony Storr
FUNDAMENTALISM Malise Ruthven
GALAXIES John Gribbin
GALILEO Stillman Drake
GAME THEORY Ken Binmore
GANDHI Bhikhu Parekh
GENIUS Andrew Robinson
GEOGRAPHY John Matthews and David Herbert
GEOPOLITICS Klaus Dodds
GERMAN LITERATURE Nicholas Boyle
GERMAN PHILOSOPHY Andrew Bowie
GLOBAL CATASTROPHES Bill McGuire
GLOBAL ECONOMIC HISTORY Robert C. Allen
GLOBAL WARMING Mark Maslin
GLOBALIZATION Manfred Steger
THE GOTHIC Nick Groom
GOVERNANCE Mark Bevir
THE GREAT DEPRESSION AND THE NEW DEAL Eric Rauchway
HABERMAS James Gordon Finlayson
HAPPINESS Daniel M. Haybron
HEGEL Peter Singer
HEIDEGGER Michael Inwood
HERODOTUS Jennifer T. Roberts
HIEROGLYPHS Penelope Wilson
HINDUISM Kim Knott
HISTORY John H. Arnold
THE HISTORY OF ASTRONOMY Michael Hoskin
THE HISTORY OF LIFE Michael Benton
THE HISTORY OF MATHEMATICS Jacqueline Stedall
THE HISTORY OF MEDICINE William Bynum
THE HISTORY OF TIME Leofranc Holford-Strevens
HIV/AIDS Alan Whiteside
HOBBES Richard Tuck
HUMAN EVOLUTION Bernard Wood
HUMAN RIGHTS Andrew Clapham
HUMANISM Stephen Law
HUME A. J. Ayer
HUMOUR Noël Carroll
THE ICE AGE Jamie Woodward
IDEOLOGY Michael Freeden
INDIAN PHILOSOPHY Sue Hamilton
INFORMATION Luciano Floridi
INNOVATION Mark Dodgson and David Gann
INTELLIGENCE Ian J. Deary
INTERNATIONAL MIGRATION Khalid Koser
INTERNATIONAL RELATIONS Paul Wilkinson
INTERNATIONAL SECURITY Christopher S. Browning
ISLAM Malise Ruthven
ISLAMIC HISTORY Adam Silverstein
ITALIAN LITERATURE Peter Hainsworth and David Robey
JESUS Richard Bauckham
JOURNALISM Ian Hargreaves
JUDAISM Norman Solomon
JUNG Anthony Stevens
KABBALAH Joseph Dan
KAFKA Ritchie Robertson
KANT Roger Scruton
KEYNES Robert Skidelsky
KIERKEGAARD Patrick Gardiner
THE KORAN Michael Cook
LANDSCAPE ARCHITECTURE Ian H. Thompson
LANDSCAPES AND GEOMORPHOLOGY Andrew Goudie and Heather Viles
LANGUAGES Stephen R. Anderson
LATE ANTIQUITY Gillian Clark

LAW Raymond Wacks
THE LAWS OF THERMODYNAMICS Peter Atkins
LEADERSHIP Keith Grint
LINCOLN Allen C. Guelzo
LINGUISTICS Peter Matthews
LITERARY THEORY Jonathan Culler
LOCKE John Dunn
LOGIC Graham Priest
MACHIAVELLI Quentin Skinner
MADNESS Andrew Scull
MAGIC Owen Davies
MAGNA CARTA Nicholas Vincent
MAGNETISM Stephen Blundell
MALTHUS Donald Winch
MANAGEMENT John Hendry
MAO Delia Davin
MARINE BIOLOGY Philip V. Mladenov
THE MARQUIS DE SADE John Phillips
MARTIN LUTHER Scott H. Hendrix
MARTYRDOM Jolyon Mitchell
MARX Peter Singer
MATHEMATICS Timothy Gowers
THE MEANING OF LIFE Terry Eagleton
MEDICAL ETHICS Tony Hope
MEDICAL LAW Charles Foster
MEDIEVAL BRITAIN John Gillingham and Ralph A. Griffiths
MEMORY Jonathan K. Foster
METAPHYSICS Stephen Mumford
MICHAEL FARADAY Frank A. J. L. James
MICROECONOMICS Avinash Dixit
MODERN ART David Cottington
MODERN CHINA Rana Mitter
MODERN FRANCE Vanessa R. Schwartz
MODERN IRELAND Senia Pašeta
MODERN JAPAN Christopher Goto-Jones
MODERN LATIN AMERICAN LITERATURE Roberto González Echevarría
MODERN WAR Richard English
MODERNISM Christopher Butler
MOLECULES Philip Ball
THE MONGOLS Morris Rossabi
MORMONISM Richard Lyman Bushman
MUHAMMAD Jonathan A. C. Brown
MULTICULTURALISM Ali Rattansi
MUSIC Nicholas Cook
MYTH Robert A. Segal
THE NAPOLEONIC WARS Mike Rapport
NATIONALISM Steven Grosby
NELSON MANDELA Elleke Boehmer
NEOLIBERALISM Manfred Steger and Ravi Roy
NETWORKS Guido Caldarelli and Michele Catanzaro
THE NEW TESTAMENT Luke Timothy Johnson
THE NEW TESTAMENT AS LITERATURE Kyle Keefer
NEWTON Robert Iliffe
NIETZSCHE Michael Tanner
NINETEENTH-CENTURY BRITAIN Christopher Harvie and H. C. G. Matthew
THE NORMAN CONQUEST George Garnett
NORTH AMERICAN INDIANS Theda Perdue and Michael D. Green
NORTHERN IRELAND Marc Mulholland
NOTHING Frank Close
NUCLEAR POWER Maxwell Irvine
NUCLEAR WEAPONS Joseph M. Siracusa
NUMBERS Peter M. Higgins
NUTRITION David Bender
OBJECTIVITY Stephen Gaukroger
THE OLD TESTAMENT Michael D. Coogan
THE ORCHESTRA D. Kern Holoman
ORGANIZATIONS Mary Jo Hatch
PAGANISM Owen Davies
THE PALESTINIAN-ISRAELI CONFLICT Martin Bunton
PARTICLE PHYSICS Frank Close
PAUL E. P. Sanders
PENTECOSTALISM William K. Kay
THE PERIODIC TABLE Eric R. Scerri
PHILOSOPHY Edward Craig
PHILOSOPHY OF LAW Raymond Wacks
PHILOSOPHY OF SCIENCE Samir Okasha
PHOTOGRAPHY Steve Edwards
PHYSICAL CHEMISTRY Peter Atkins
PLAGUE Paul Slack
PLANETS David A. Rothery

PLANTS Timothy Walker
PLATO Julia Annas
POLITICAL PHILOSOPHY David Miller
POLITICS Kenneth Minogue
POSTCOLONIALISM Robert Young
POSTMODERNISM Christopher Butler
POSTSTRUCTURALISM
Catherine Belsey
PREHISTORY Chris Gosden
PRESOCRATIC PHILOSOPHY
Catherine Osborne
PRIVACY Raymond Wacks
PROBABILITY John Haigh
PROGRESSIVISM Walter Nugent
PROTESTANTISM Mark A. Noll
PSYCHIATRY Tom Burns
PSYCHOLOGY Gillian Butler and
Freda McManus
PURITANISM Francis J. Bremer
THE QUAKERS Pink Dandelion
QUANTUM THEORY
John Polkinghorne
RACISM Ali Rattansi
RADIOACTIVITY Claudio Tuniz
RASTAFARI Ennis B. Edmonds
THE REAGAN REVOLUTION Gil Troy
REALITY Jan Westerhoff
THE REFORMATION Peter Marshall
RELATIVITY Russell Stannard
RELIGION IN AMERICA Timothy Beal
THE RENAISSANCE Jerry Brotton
RENAISSANCE ART
Geraldine A. Johnson
REVOLUTIONS Jack A. Goldstone
RHETORIC Richard Toye
RISK Baruch Fischhoff and John Kadvany
RIVERS Nick Middleton
ROBOTICS Alan Winfield
ROMAN BRITAIN Peter Salway
THE ROMAN EMPIRE
Christopher Kelly
THE ROMAN REPUBLIC
David M. Gwynn
ROMANTICISM Michael Ferber
ROUSSEAU Robert Wokler
RUSSELL A. C. Grayling
RUSSIAN HISTORY Geoffrey Hosking
RUSSIAN LITERATURE Catriona Kelly
THE RUSSIAN REVOLUTION
S. A. Smith
SCHIZOPHRENIA Chris Frith and
Eve Johnstone
SCHOPENHAUER
Christopher Janaway
SCIENCE AND RELIGION
Thomas Dixon
SCIENCE FICTION David Seed
THE SCIENTIFIC REVOLUTION
Lawrence M. Principe
SCOTLAND Rab Houston
SEXUALITY Véronique Mottier
SHAKESPEARE Germaine Greer
SIKHISM Eleanor Nesbitt
THE SILK ROAD James A. Millward
SLEEP Steven W. Lockley and
Russell G. Foster
SOCIAL AND CULTURAL
ANTHROPOLOGY
John Monaghan and Peter Just
SOCIALISM Michael Newman
SOCIOLINGUISTICS John Edwards
SOCIOLOGY Steve Bruce
SOCRATES C. C. W. Taylor
THE SOVIET UNION Stephen Lovell
THE SPANISH CIVIL WAR
Helen Graham
SPANISH LITERATURE Jo Labanyi
SPINOZA Roger Scruton
SPIRITUALITY Philip Sheldrake
STARS Andrew King
STATISTICS David J. Hand
STEM CELLS Jonathan Slack
STUART BRITAIN John Morrill
SUPERCONDUCTIVITY
Stephen Blundell
SYMMETRY Ian Stewart
TEETH Peter S. Ungar
TERRORISM Charles Townshend
THEOLOGY David F. Ford
THOMAS AQUINAS Fergus Kerr
THOUGHT Tim Bayne
TIBETAN BUDDHISM
Matthew T. Kapstein
TOCQUEVILLE Harvey C. Mansfield
TRAGEDY Adrian Poole
THE TROJAN WAR Eric H. Cline
TRUST Katherine Hawley
THE TUDORS John Guy
TWENTIETH-CENTURY BRITAIN
Kenneth O. Morgan

THE UNITED NATIONS Jussi M. Hanhimäki
THE U.S. CONGRESS Donald A. Ritchie
THE U.S. SUPREME COURT Linda Greenhouse
UTOPIANISM Lyman Tower Sargent
THE VIKINGS Julian Richards
VIRUSES Dorothy H. Crawford
WITCHCRAFT Malcolm Gaskill
WITTGENSTEIN A. C. Grayling
WORK Stephen Fineman
WORLD MUSIC Philip Bohlman
THE WORLD TRADE ORGANIZATION Amrita Narlikar
WRITING AND SCRIPT Andrew Robinson

Available soon:

ALEXANDER THE GREAT Hugh Bowden
COMPLEXITY John Holland
GENES Jonathan Slack
GOD John Bowker
HORMONES Martin Luck

For more information visit our website
www.oup.com/vsi/

Charles Sheppard

CORAL REEFS

A Very Short Introduction

OXFORD
UNIVERSITY PRESS

Great Clarendon Street, Oxford, OX2 6DP,
United Kingdom

Oxford University Press is a department of the University of Oxford. It furthers the University's objective of excellence in research, scholarship, and education by publishing worldwide. Oxford is a registered trade mark of Oxford University Press in the UK and in certain other countries

First edition published 2014

Impression: 1

Published in the United States of America by Oxford University Press
198 Madison Avenue, New York, NY 10016, United States of America

British Library Cataloguing in Publication Data
Data available

Library of Congress Control Number: 2014938179

ISBN 978–0–19–968277–5

Printed in Great Britain by
Ashford Colour Press Ltd, Gosport, Hampshire

Contents

List of illustrations

Under water photographs by Anne Sheppard and Charles Sheppard

Chapter 1
Geology or biology?

Discovery and early explorers

Coral reefs have been regarded with awe for centuries. Early seafarers were wary of them, naturalists were confused by them, but many coastal people benefited greatly from these mysterious rocky structures that grew up to the surface of the sea. They have been the source of a rich supply of food for coastal people, and they provide a breakwater from storms and high waves to countless coastal communities that developed in their lee.

Long before scientists started to wonder what coral reefs really were, they had been encountered and used in many ways. Awareness of reefs beyond the local villages that used them for food started, perhaps, amongst pilgrims sailing down the Red Sea to Jeddah, the port serving the holy city of Mecca, for that part of the Red Sea is one area that is gloriously rich—or dangerously strewn, from a seafarer's point of view—with coral reefs. And then, from a European perspective, as trade developed to both the East and West Indies, coral reefs were increasingly encountered. They were a navigational hazard that provided fatal resting places for countless traders because their biology, not understood until much later, leads them to grow up to the surface of the sea at low tide, just where they offer the greatest danger to sailing ships. Perhaps the most famous European explorer to suffer the consequences was Captain Cook, who nearly came to an early demise on the Great Barrier Reef as he explored the coast of Australia.

Nevertheless, the value of coral reefs has always far exceeded the hazards they pose to navigation. Their scale is enormous; the Indo-Pacific coral reef province, for a start, is the largest ecosystem on Earth, and probably its most diverse, at least at higher taxonomic levels. Ever since mankind spread out from Africa, spreading eastwards through the Indonesian and Philippine archipelagos and onwards to the Pacific Islands, people have inhabited land near reefs or islands on top of them, attracted by the huge quantities of fish and other marine life that were available for food. People living on small islands or beside a coral coast commonly obtained almost all their protein from fishing or gleaning at low tide, and many still do. Even those who lived a little further from the water would obtain a significant proportion of their food from those who caught more from the reef than they needed, in a proportion diminishing with distance from the rich source.

Europeans mostly travelled west across the Atlantic at first, where they encountered the coral fringed islands of the Caribbean. Columbus's first New World landfall was made tricky by the coral reefs of San Salvador island in the Bahamas (some say), which the intrepid navigator thought was Cathay on the other side of the world. Following the early days of exploration to the West Indies came the age of naturalists. At this time, the coral island of Bermuda, which lies to the north of the main Caribbean region, became important in providing these early exploring naturalists with a half-way stop on their journey. This period of exploration promoted the beginnings of modern scientific understanding, because many of the ships of discovery also carried a naturalist as a passenger, possibly appointed by a king or an archduke on whose patronages so much early discovery depended. Slightly later, ships bearing these naturalists headed eastwards too, sometimes on journeys of exploration, but more commonly as appointed passengers. One naturalist, Carsten Niebuhr, wrote of coral reefs as he travelled east:

> I have already had occasion to speak, in the course of my travels, of the astonishing mats of works formed by the marine insects; namely the immense banks of coral bordering, and almost filling up [the sea]. The reader may therefore conceive with himself what a variety of madrepores and millepores are to be met with in these seas.

He wrote this in 1792. There was in those days, and for a long time to come, an almost complete inability to see or understand what went on beneath the surface of the water. Sea monsters still decorated nautical charts, and the monsters were only slightly less useful than the positions marked of many of the reefs. Fifty years later, in 1840, also travelling eastwards, J. Raymond Wellstead was still mystified by it all and wrote:

> there are secrets on the surface as well as within the bosom of the ocean, which lie shrouded from human observation and research ...where there is mystery there will always be interest, and to the greater the one, the more intense the other.

Another naturalist of that time, C. G. Ehrenberg, who became very well known in coral science circles, wrote in 1832:

> hummingbirds sport around the plants of the tropics, so also small fishes, scarcely an inch in length and never growing larger, but resplendent with gold, silver, purple and azure, sport around the flower-like corals.

These naturalists were clearly fascinated by coral reefs and probably were equally frustrated by not being able to see them properly. But some were starting to get closer to them; another half century later, in 1878, C. B. Klunzinger wrote:

> Nowhere can one contemplate the life of the corals, and what belongs to it, more quietly and comfortably than here, although he has to lie on his belly—a trifling matter for the naturalist—and hold his magnifying glass at the point of his nose above a coral bush.

Meanwhile Alfred Russel Wallace, the co-discoverer of evolution, wrote of Indonesian reefs in 1869:

> The clearness of the water afforded me one of the most astonishing and beautiful sites I have ever beheld. The bottom was absolutely hidden by a continuous series of corals, sponges, actinae, and other marine productions, of magnificent dimensions, varied forms and brilliant colours…it was a sight to gaze at for hours, and no description can do justice to its surprising beauty and interest.

And this amazement came after he had explored such exotic parts of the world as the Amazon, not to mention Indonesia's islands.

Explanation of how reefs grow—the 'Coral Reef Question'

In the 19th century geologists were the leaders in research on coral reefs, developing an understanding of how these remarkable structures developed. Considerable controversy and impassioned arguments developed around how reef structures grew upwards to sea level and then stopped. In one sense this was easy: they were somehow produced by marine life and everybody knew that marine life would not grow out of the water. But what were they? What were the animals—or were they plants as was thought for quite a while—and how did they make the rock? The term 'animacule' was coined at one stage to describe the baffling small bunches of waving tentacles that were attached to the top of the rocks they sat upon, some a few inches long, but mostly very tiny. It was clearly very confusing at the biological level, though at the geological level the early scientists were starting at least to see what had actually happened.

Scientists knew by then that this rock was limestone, calcium carbonate, and a very pure form of it. They knew too that the shells of many animals were also made of limestone. But although it was well known that marine organisms produce limestone, these reef structures were immense. Inland, many miles from the sea, were other limestone hills and mountains, clearly very ancient. Could marine organisms have built these too? They certainly contained marine fossils, which suggested that indeed they had—that is, unless the naturalist had a biblical view of time spans, which certainly posed an unsolvable problem to quite a number of these naturalists. Before the full recognition of an ancient Earth, the implied vast timescales served to baffle as much as to enlighten.

Charles Darwin's famous voyage on the *Beagle* was one of the turning points in coral reef science. Although better known for his ideas on evolution, I think his 1842 book on coral reef formation and large-scale movements of parts of the Earth's surface were more innovative and original. His theory depended on the massive subsidence of volcanoes (and other parts of the Earth's surface). Darwin proposed that coral animals could grow only in the surface regions, and as the land against which they grew subsided, the corals kept growing upwards, on top of each other, to maintain their position in shallow and warm waters. It had already been observed, in sections taken from fossil reefs, that corals were embedded in the limestone matrix, so this was not the shock; what surprised and was contested by some was the huge scale of space and time needed, if Darwin's explanation was correct. His theory had a simple but powerful elegance along with its explanatory power, and it eventually became accepted and then proved long after his own lifetime, as we shall see.

But in Darwin's time this was still only a partial theory, and it did not displace all the earlier theories for many years. For example, in

1821 Von Chamisso had noticed that reef-building corals grow best in turbulent waters that are found on the outer periphery of most reefs. This led him to believe that in such conditions corals would grow upwards so that, whatever their foundation, they would reach the surface in the annular shape typical of a coral atoll. Two famous collaborators, Quoy and Gaimard, opined that corals simply grew up around the rim of old volcanoes. The coincidence of so many volcanoes happening to grow to identical height just below the water surface was never explained. Nevertheless, the latter theory was even supported by the great geologist Charles Lyell until he dropped it in favour of Darwin's theory.

The accepted mechanism today is this: before any subsidence event, for example around the flanks of a new volcano—anywhere on the flank, not specifically the rim—corals would establish in the photic zone (i.e. where there is enough light). The volcano would then sink a little and corals would grow up, fringing the island and forming what we call a fringing reef (Figure 1). Then, with further subsidence of the volcano and upward growth of the reef, the corals appear (to the observer on the surface) to detach from the land and to form a reef that is increasingly distant from shore—a barrier reef. One definition of a barrier reef has been simply that there should be a navigable passage between reef and the shoreline.

The final stage of this sequence is that further subsidence submerges the volcanic island completely so that all that is now visible from the surface is a ring of coral which is still maintaining its growth up to the surface: the coral atoll. It is not the case that a ring of coral is simply sitting on top of the rim of an old volcano—this is far too simplistic for many reasons. For one thing the old volcano might be hundreds or thousands of metres deeper, and we now also know that repeated lowering and rising of sea level with glacial cycles has greatly changed and modified any such simple structure, and that there have been repeated episodes of growth interlaced with periods of no growth and erosion. Added to that, limestone is relatively easily dissolved by rainwater

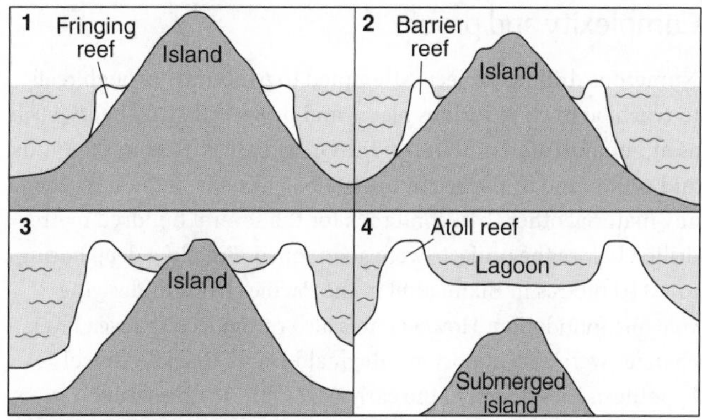

1. Sketch of Darwin's scheme of the sequence of progression of a fringing reef, barrier reef, and atoll as its central volcano subsides over millions of years

because it is an alkaline rock and rainwater is always slightly acidic. This means that substantial erosion of old reef above sea level has also occurred during prolonged periods when sea level dropped in one or other of several geological periods. All these add to the complexity of what we will see at the surface today.

Darwin's broad theory of atoll formation by subsidence is generic and, while accurate enough, it only partially explains the formation of so many of the world's complicated networks of barriers, islands, and fringing reefs, especially in shallow areas such as those found in much of the Indonesian region, for example. The many exceptions and complications to Darwin's simple progression led to more than a hundred years of further debate, showing that this was a complex problem requiring more than one explanation. Coral reefs will grow upwards to meet the surface when sea level is high, and will be eroded away during periods when sea level is low. Progressive understanding of not only vertical land movement but of vast changes in sea level lasting thousands of years, usually during ice ages, showed the layers of complexity that superimpose Darwin's central, but basically correct, idea.

Complexity and proof

Numerous drilling projects attempted to penetrate through reefs to reach bedrock in various places and prove Darwin's theory, such as at Funafuti atoll in 1898, on the Great Barrier Reef in the 1920s and 1930s, and in Borneo in the 1940s, but none of these reached any material other than coral rock for the several hundred metres drilled before the projects were abandoned. Finally, a deep boring to 1,340 metres in Bikini atoll in the Pacific struck underlying volcanic foundation. However, it is also recognized that sea level changes were also important. Reginald Daly's Glacial Control hypothesis, developed in the early part of the 20th century, showed how great changes in sea level were crucial too. At the time, many viewed it as a challenge to Darwin subsidence theory, but in reality it simply complements it. David Stoddart wrote: 'If Darwin's was a theory of reef *structure*, Daly's was one of reef *surface morphology*.' Erosion has had a great effect on the present topography of reefs, particularly in areas of the world with relatively rapid land movements.

Several other ideas were then developed to explain differences seen in various places. The antecedent platform theory was developed to explain why barrier reefs and atolls can develop on any suitable submerged platform, provided it is at a depth reached by sunlight, and that fringing reefs need not form an intermediate stage of the process. This theory required vast numbers of similarly located platforms, and an explanation for this unlikely event included wave erosion, which might reduce all volcanoes to the same shallow depth. This theory emphasized the importance of erosion, making it as important as the growth of corals. It was known, too, that erosion of old reef by rain results in saucer-shaped depressions in large areas of limestone, which creates atoll-like dishes. This karstic-saucer theory developed by Edward Purdy certainly explains many structures within reefs, such as the deep cavities in reef rock known as 'blue holes'.

As with many areas of research, scientists often tend to have a preference for one single theory, being unable to accept that different theories may be plausible, depending on place, scenario, and time. This position taken by scientists in coral reef research has been seen throughout its history, including in the present context relating to what is causing today's demise of coral reefs generally. Scientists may be strongly influenced in their preference of focus by where in the world they have worked and which reefs they have observed. In fact, all these processes with theories relating to them take place to a greater or lesser degree in different places. But the possibility of a scientist visiting sufficiently numerous and varied locations to see all possible examples of reefs the world has to offer is difficult even today, and was simply not possible in the days of sailing ships.

Biological discovery

While these geological explanations were being developed, the reef's component corals, fish, and other groups were being described extensively by scientists (Figure 2), though perhaps with limited understanding. Not until about three-quarters of a century after the early geologists' explanations did biological scientists start to understand life on the reefs in an ecological way, to discover how the reefs worked, how their components fitted together, and what their most important components were. The colours and apparent profligacy of life on reefs had long attracted naturalists who were much more used to life on the shores of their own, usually colder, countries. However, an ecological understanding of function principally followed the advent of diving apparatus, even if some important progress had been made through the use of microscopes and experiments in laboratories and on very shallow reefs.

Most of this book is about the biology and ecology of reefs that has been discovered in the past few decades. The obvious biological complexity of reefs makes them rather daunting to understand.

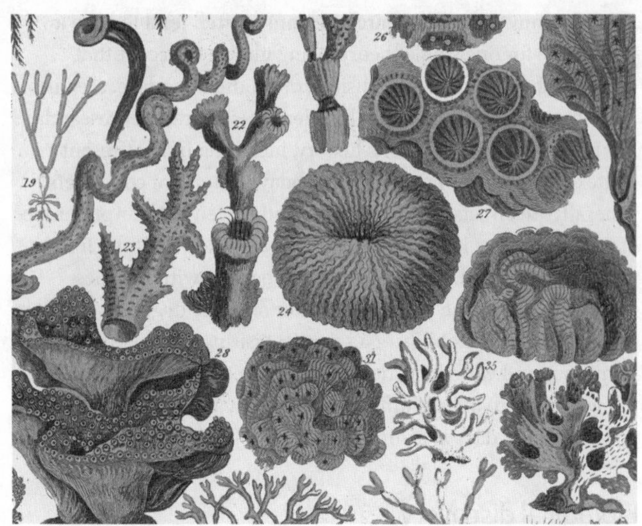

2. Early taxonomic drawings of corals were very accurate and detailed, though they were usually done from fragments that were dredged up and sometimes rather damaged in the process

They support more kinds of life than any other area in the sea, and reefs themselves are built by that life. Their complexity is matched by their importance: they show enormously high rates of protein production—when they are not damaged by too much exploitation. This being the case, they are increasingly important to feeding the growing number of people that demand (what economists now call) their 'services'. A biologist might not much like reducing the idea of a reef to something that simply provides food, a mere larder with monetary value, but in this overcrowded and overexploited world that is increasingly the way of society.

To many scientists, reefs are the most intriguing ecosystems that have ever appeared in the world's oceans. They contain many scientific paradoxes: in some ways they are robust, but in others they are very fragile. They contain a riot of species which appear to be visually chaotic, and the reasons for the observed vibrancy, colour, and diversity are being better understood all the time. Coral reefs have provided material for greater insights into how nature is organized. It is hoped that we will find a way to interact with and use reefs with a minimum of disruptive impact. In addition, we should even be able to please the economists and governments of the world by showing how reefs can continue to provide their services to the burgeoning human population.

Chapter 2
Ancient reefs and islands

Ancient reefs

To keep the reef story in perspective we should not think of the present coral reefs as being automatically a permanent fixture of the Earth. In the past, several other groups of organisms formed reefs for hundreds of millions of years but then became extinct. We have long been aware of many ancient formations built from a wide range of organisms. There is nothing new about massive, limestone-deposited structures on our planet, uplifted to form hills and mountains.

In Precambrian times, assemblages of simple marine organisms laid down reefs over aeons of time. These were stromatolites, built by photosynthetic cyanobacteria, depositing limestone but also trapping sediment to form cauliflower-like layered domes. This was over three billion years ago, in the Archaean Eon, when atmospheric oxygen was low, long before the advent of complex life.

Atmospheric oxygen increased significantly from about 1.8 billion years ago, a change that permitted the development of more complex, multicellular, oxygen-using forms of life. Existing reefs declined but around the beginning of the Cambrian Period a new group of reef-building organisms called Archaeocyathids, small sponge-like organisms that were probably filter feeders, developed to reach their peak, which soon saw an explosion of animal life, as documented by the exquisite fossils of the Burgess Shale and Cheng Jiang biota. The archaeocyathids became extinct at the end of the Cambrian when, for a few more tens of millions of years, stromatolites were again the main reef-builders. Around that time sponges, distant ancestors of some of today's sponges, became more important as reef-builders. Sponges never went away, and are still abundant on reefs. These were joined for many millions of years by stromatoporoids, which appear to be related to sponges, and which dominated from the Ordovician Period but began to decline by the time that the plant matter making today's great seams of coal were being deposited, in the Carboniferous.

Two groups of corals then began to form reefs, but these were not the ancestors of today's corals. The first group was the tabulate corals, which were almost entirely colonial, and the rugose corals which existed in colonial and solitary forms. Both groups lasted until the end of the Permian (about 250 million years ago), and during this long span they built a variety of reef types ranging from small patch reefs to extensive barriers. All this came to an end at the great Permo-Triassic extinction event. This extinction had a massive effect on the world's life, seeing the demise of perhaps over 90 per cent of marine species. As Raup put it in 1998: 'if these estimates are even reasonably accurate, global biology (for higher organisms at least) had an extremely close brush with total destruction'. Even sponges, which thrive today, came close to extinction. Several causes have been suggested for this extinction, but a likely possibility is that the climate changed, a salutary point in today's rapidly changing environment.

The main reef-building animals of today, the scleractinian corals, came to dominance about 240 million years ago. These are cnidarians, but what exactly they developed from is debated; they may have arisen from anemone-like relatives, or may be ancestral to them, and some have suggested that different groups or families of corals may have different origins. Corals do not appear in the fossil record until well after the great Permian-Triassic extinction event, although that geological interval has been very well investigated. They probably developed from relatives, like anemones. Their biology is described in greater detail in Chapter 3, but the fact that they became so successful as reef-builders is because they were able to form symbiotic associations with single celled algae called Zooxanthellae, so that by the middle Triassic they were forming substantial limestone structures. This symbiosis is a crucial aspect of coral reef life.

The next great extinction occurred at the end of the Cretaceous Period. The so-called K-T extinction, which removed the dinosaurs, killed about 70 per cent of corals, and greatly suppressed reef-building for a long time, but they then returned with vigour.

The deposition of limestone in various forms has been a characteristic of all these reef-forming groups, and perhaps its main significance is that they all could both create their own substrate and modify their environment to a considerable extent. The deposition of limestone requires energy and certain conditions to be met in terms of carbon-dioxide–carbonate balance, temperature, and other physical factors. The growth of these structures is not passive, therefore, and reef-building activity requires fairly tight environmental parameters. Since the earliest days of corals, atmospheric and oceanographic conditions on Earth have varied, often substantially, but corals have thrived throughout this time, forming the antecedents for today's coral reefs.

This history is important. Present day ecology, the organization of species on a reef and their behaviour—basically, how a reef *works*—is a product of this long ancestry. Climate has fluctuated greatly throughout this time; sea levels have changed, and the temperatures of the Earth have changed too. The riot of life seen on healthy reefs today is the result of a very long heritage. The processes are complex, as would be expected from what is probably the most diverse environment that the Earth has seen in all its biological history.

Coral islands—tips of the icebergs

Coral islands have long been a key component of coral reefs and the target of research. They are commonly the first parts of reefs that are seen. They are made from reefs, and sit on top of them. Home to millions of people, they come in different forms, but in terms of their structure there are really only a few different kinds, with many islands showing combinations of all of them. The simplest are the true coral cays. These range from being little more than a sand bank to larger islands, richly clothed with vegetation. They may form when sand, which is a product of a broken down and eroded coral and reef matrix, gets pushed onto the top of a reef by wave action. If enough sand is deposited, birds may settle. The birds leave guano, which in turn fertilizes seeds that have arrived by wind and birds; some plants take root, precariously at this stage; and this attracts more birds, who leave more guano. As more sand gets pumped onto this now dry platform, plant roots stabilize it and prevent it from being immediately washed away in a storm. In time a coral cay is born. Then, with help from microbes and various chemical reactions, the sand grains become cemented together to form solid rock. Details of how this happens remains poorly understood, but the result is a small island (Figure 3).

A more complicated variant in coral island formation can be recognized immediately if it contains cliffs that surround it, especially when these rise several metres above sea level. These islands have solid blocks of limestone at their core, which are made up of much older reefs or limestone sand dunes, formed when sea levels were higher than they are now, before or during

3. Three small islands in the central Indian Ocean. They all have a core of old coral reef, some parts being about 8,000 years old, but the larger two are now covered with recently formed sand and rubble. The foreground island has an extensive and complicated reef flat, and there is much less reef flat around the furthest one. Note also the bite out of the bottom right of the nearest island, where rising sea levels and reef mortality is causing attrition of the land

the last ice age. At the maximum extent of glaciers it is thought that these blocks protruded above the present surface of the sea by many tens of metres, and indeed such tall limestone stacks can be seen protruding from the sea today in several parts of the world. A gradual rise of the sea level starting about 18,000 years ago, as well as erosion of the limestone rock from rain, which is always slightly acidic, has left emergent lumps of rock protruding just a few or a few tens of metres above the surface of the sea. Their core may be a mixture of consolidated limestone rubble and rock. These limestone stacks will usually have shifting areas of sand around the edges, but their altitude and solid core gives away the fact the hearts of these islands are of much older limestone.

Carbon dating shows that they can be anywhere between 6,000 and several hundred thousand years old. Many islands in the Caribbean are of this kind.

Popular with tourists are what are known as high islands, which are basically volcanic or basaltic islands fringed with coral reef. Many islands in the Pacific and Indian Oceans, and large numbers in the Caribbean are of this kind—they are ancient or even currently active volcanoes, and other land-forms which may still be subsiding into the sea. They may have lush interiors, if they benefit from high rainfall, or they may be more barren, for example in parts of the tropics where the winds do not bear much moisture. Extending this to larger landmasses, one can see all variants of this type of island in many countries, often all within one visual sweep of the landscape. The coralline or reefal component of some islands may be relatively small compared to other rock.

The result is countless coral islands around the tropical world. Many entire nations are built exclusively on coral reef islands and it is salutary to remember that a significant number of states in the United Nations General Assembly are there only because of corals and reef processes. As we will see later, many of these countries are in trouble because of the problems arising from being only marginally above sea level, and being dependent on the living processes of the reefs, this makes them potentially very vulnerable.

Rain is a great structuring force on these islands, both in its amount and its frequency. Rain brings life, and the amount and kinds of vegetation on the coral islands are hugely affected by the interval between rain periods as well as total quantity of rain. In those parts of the ocean where there is very little rain, coral islands may never have supported very much life, and may never have been successfully colonized or developed—into coconut plantations, for example. The substrate after all is very porous and fresh water may not remain in the soil for very long. In other places, heavy rainfall may last for several months each year and much fresh water penetrates the soil and porous rock of the island, forming lenses beneath the land surface, slightly above the level of the sea. It is this life-giving water table that keeps seawater at bay, and which supports island vegetation. While rain is plentiful, island life will thrive, commonly those species that show some tolerance to salt. Where this is the case, human communities inevitably have developed. These islands may be an attractive destination today for tourists too; reef-based tourism in some coral island nations can form 40 to over 80 per cent of their entire national income. Even more importantly perhaps, the reefs provide the people with huge quantities of food.

Today's reef crisis

But we are changing the environment around these coral reefs, which is having severe consequences. In later chapters, I will try to show why the present changes to ocean temperature and acidity mean that, at current rates of change, the present rate of coral reef formation simply cannot continue for much longer. Increasing numbers of scientists have given increasingly urgent warnings, but what can be done about it? Coral reefs as a whole have been described as being the canaries in the coal mine—they are the part of the ecosystem that will be the first to succumb to overexploitation and abuse.

Since our direct observations started, with the introduction of diving equipment, we have had just 40 to 50 years of fruitful coral reef research. During this time, several thousand reef researchers have been advancing our understanding of how these structures function. While this appears to be a large number of researchers, it is in fact no more than the number that may be found at a single medical or business conference at a single time. Neither is it such an impressive number when we consider how many people in the world depend on reefs for food, land, or revenue. Unfortunately and based on present trends, this focus of biological research might be short-lived. In this time, marine scientists have frequently observed, worried, and issued warnings about many aspects of degradation of coral reefs in all our oceans, Caribbean, Indian, and Pacific, especially those in some of the more densely populated and heavily utilized areas.

Reefs today, so soon after the start of their 'modern' period of scientific flourishing, are dying. About a quarter have already died and another half have degraded. Yet although scientists' warnings have grown increasingly grim, those in authority have continued essentially to ignore them. Take the Caribbean, for example. In the 1970s the shallowest three metres of just about all reefs there were literally covered with the world's largest species of coral, *Acropora palmata*, known as elkhorn coral. At low tide you simply could not pass across such a reef into the deeper water beyond because of the presence of these dense forests of branches that grow as tall as a person. Spectacular in the extreme, no diving magazine would miss an opportunity to print new photos of them. But, by the 1980s something called White Band disease had killed almost all of them. Over the years, these huge coral skeletons crumbled, so that by the 1990s most of these glorious forests were rubble. The three dimensional reef structures had become two dimensional, without much life and no longer strong enough to break the waves and protect shorelines. Not quite all the elkhorn died—some have remained and can be seen today, with possible evidence that in some places they are recovering. However, nowhere are there high densities of elkhorn remaining—the entire shallowest portion of the Caribbean reef ecosystem has essentially disappeared. It is the same story elsewhere: in the coral-rich Gulf of Aqaba I used to observe uncountable numbers of table corals, also of the genus *Acropora*, but during a visit to Jordan 20 years later at the turn of the century, I was taken, proudly and touchingly, to view the only table coral thought to be left in Jordan at that time. In this case, global warming was the probable cause of mortality, and heat-susceptible table corals had been greatly reduced throughout the northern Red Sea at that time.

What caused this mortality, which happened so soon after the start of the great age of coral reef discovery? That is explained in the final third of this book. For those who saw these reef ecosystems before their degradation, or for those fortunate to see some of the remaining reefs in truly good condition, the decline, as can have been witnessed in a single observer's lifetime, has been as scientifically spectacular as it has been alarming. As such, coral reefs offer us an important warning. We have been ignoring the health and welfare of an ecosystem that supports us, one which has brought pleasure, food, and value to so many—coral reefs offer us much more than a tale of a wonderfully complex ecosystem. They offer us a mirror, or a microcosm perhaps, of humanity as a whole.

Chapter 3
The architects of a reef

A diver seeing his or her first coral reef may be visually overwhelmed. Hundreds of coral species and thousands of colourful fishes and other species—a near infinity of colours, shapes, and sizes—confuse the eye. Fish add tremendous activity, singly, in pairs, or in schools, feeding on bottom-dwelling life and on each other (Figure 4). Sea fans as tall as the diver flex as each wave passes overhead, and crackling noises made by reef creatures fill the sea with sound.

There is structure to this kaleidoscope of course, acting at many scales. There is a pattern across the world (Figure 5), a pattern across a single reef (with many similarities and consistencies between most reefs), and a pattern in the small area that a diver can see at any one time. There is a common underlying biological form in much of what we can see. This chapter gives a very brief global overview of the distribution of reefs, and then details the biology of the main animals and plants that build them.

The global pattern

Coral reefs are tropical ecosystems. While all reefs contain a very high diversity of life, different areas of the coral seas support large differences in levels of diversity in groups of animals and plants. Southeast Asia's is especially diverse while that of some areas such

4. A thriving reef slope in a no-fishing marine reserve. Fish frequently obstruct the view of the rich corals

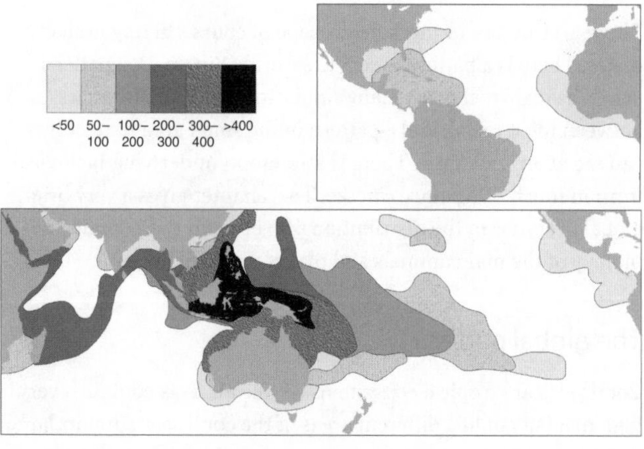

5. Map showing the distribution of reef building corals and their diversity. The 'coral triangle' is in black, in the Southeast Asian region

as the Caribbean and Eastern Pacific have comparatively low diversity. However, reefs in these last two areas are every bit as substantial as those in higher diversity areas because high diversity does not necessarily equate with good or strong reef construction. Indeed, in some richly populated areas, corals may be unable to create biogenic reefs at all; typically, the corals may grow on a rock, usually igneous, from which they readily detach after death without any build-up of limestone, yet the associated sea bed and fish fauna will be those of a coral reef.

The high diversity of Southeast Asia's reef and the steady decline in diversity as one moves further away from it have long been recognized but are still not well explained. Diversity broadly decreases eastwards and westwards from this region. There is also a marked decline in diversity of most taxonomic groups along a latitudinal gradient too, being highest nearest to the equator and falling towards cooler water. This is seen very clearly in the decline in diversity of coral species as one moves southwards along the Great Barrier Reef, from well over 300 species in each of the northern and central regions to 244 in the southern region, and just 87 at Lord Howe Island, which is at 32° South. Similarly, over 50 species of corals can be found in the Caribbean, but only 21 are known from Bermuda which, at 32° North is the most northerly reef system. Importantly however, the vibrancy and productivity of a reef appears to be more or less unrelated to the level of diversity it supports.

Set within this global latitudinal and longitudinal pattern are localized patterns resulting from regional environmental characteristics. Some regions may be particularly rich, for example the Red Sea, which appears to trap larvae from the Indian Ocean as well as having many endemic species. In other regions coral species may struggle to grow because of factors such as increased amounts of dissolved nutrients derived from river discharges or oceanic upwelling systems. Salinity and temperature extremes constrain corals, as does the remoteness of the region and the corals' position relative to the prevailing currents that affect it from other richer areas. The reef region with the poorest diversity is that of the southern Atlantic, where Brazil has a few reefs in restricted areas. These reefs contain just 19 coral species because of oceanographic isolation from the Caribbean due to the massive fresh water barrier caused by the Amazon and Orinoco rivers. Tropical West Africa has some corals, but these fail to form reefs at all because of the highly turbid waters there. Such regional isolation may lead to a lower diversity but also to a high degree of endemism.

Coral animals—the dominant architects

The dominant groups of biota on coral reefs are the corals themselves, which are coelenterate animals, mostly living in colonies but with some solitary species. All lay down a skeleton of limestone. The proportion of each species or group appears to depend, amongst other things, on water quality, exposure, and depth of the site.

Coral reefs are not a single, homogenous ecosystem, rather they support, and are built by, many different assemblages of corals. Think of forests of trees: there are pine forests, oak forests, beech forests, and many others, each containing entirely different dominant and structural species; some are virtually monocultures (like in a pine forest) and some have much higher diversity (e.g. in a tropical rainforest). With reefs too, there are many kinds formed by different combinations of species, and the main architectural forms in each case can differ markedly. In the Indo-pacific, branching forms of the genus *Acropora* dominate in many areas, leafy forms of several genera dominate deep regions, and huge boulders several metres across of the genus *Porites* dominate in middle depths of sheltered reefs. Some genera of many species such as *Montipora* are relatively inconspicuous but are crucial in reef-building and have members that occupy a wide range of habitats. And, like forests, one dominant form may grade into another. Yet we still think in terms of 'a coral reef', and this has, as we will see later, masked some important changes that are going on today. However, we are getting ahead of ourselves; coral reefs of all kinds, like forests, have many unifying characteristics that apply wherever they may be found. And although corals are, unsurprisingly, key to any coral reef, they do not and cannot act alone.

Coral biology and growth

Corals are anthozoans, a word derived from the Greek words 'flower' and 'animal', which describes the appearance of each polyp. Each coral colony consists of numerous small polyps, each with a ring of six (or multiples of six) tentacles surrounding a mouth, leading to the main body cavity. The polyps' tissues deposit beneath themselves calcium carbonate in a crystalline form called aragonite. A typical polyp secretes and occupies a limestone cup called a corallite, which grows upwards and outwards. The fleshy polyps generally emerge only at night, retracting into the corallite in daylight, though as with anything biological there are several exceptions.

Some species contain one solitary polyp, but in most species the polyps divide at intervals into two or more daughter polyps that continue to grow upwards until they also divide. Figure 6 shows examples of several variations on this. In many forms, each polyp remains connected to adjacent polyps via thin tissue that is also capable of depositing limestone beneath it (left side, Figure 6). Where this is the case, the limestone between the polyps continues to grow upwards along with the polyps. In many cases, when polyps bud they pinch off to form separated polyps with no connecting tissue (bottom right, Figure 6). In others, chains develop, which are in one sense a mixture of the two kinds (top right, Figure 6). Repeated patterns of growing, budding, and pinching off lead to a great variety of colony form: domes, branching thickets, table shapes, columns, and leaves. It matters

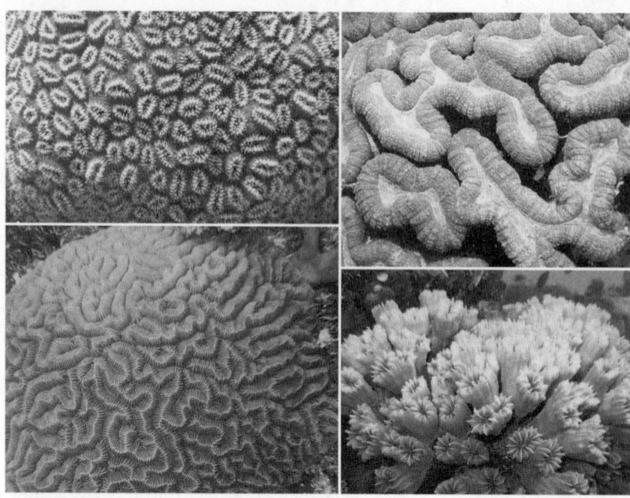

6. Examples of coral shapes. Top left, polyps are in cups, joined by live tissue (Carribean *Dichocoenia*). Bottom left: incomplete polyp budding, so series of connected polyps run along each valley (Caribbean *Meandrina*). Top right. Chains of polyps again, but each series is disconnected from neighbouring series (Indo-Pacific *Lobophyllia*). Bottom right, polyps essentially disconnected from all others, and some cups are budding into pairs (Caribbean *Eusmilia*)

also whether the daughter polyp divides from its parent from within the ring of tentacles or from lower down on the body stalk of the parent: each will lead to different shapes of the ever-growing colony. In species in which tissue continues to connect all the polyps, a touch on one side can lead to a slow ripple of colony retraction across the rest of the coral, showing that a neural network extends between polyps. In others, such neural connections are lost as the colony grows.

The variety in these structures is used to distinguish between species. The problem with this is that colonies of any one species can vary in shape a great deal according to water and light conditions. It only requires a tiny variation in one aspect, for example the amount of growth that happens between polyp budding events, to lead, over many repeated occurrences, to strikingly different colony shapes.

The key feature as far as reef-building is concerned is that the coral tissue lays down limestone. Corals thus create their own substrate, and it is this process that ultimately creates reefs.

Coral reproduction and larvae

Reproduction in corals by means of budding is asexual, and in one colony all polyps are clones—genetically identical. Asexual reproduction is also used by many species to disperse their colonies, and this is common amongst branching species that fragment to become dispersed by wave action. This may be the main method of colony propagation for some species and, in this way, lagoonal areas may become covered by thousands of square metres of, say, branching staghorn coral. Other variations occur too, an important one in lagoonal areas being that shown by the long polyped *Goniopora stoksi*; this develops 'polyp balls', which are small spheres of skeleton that develop within the fleshy mass of polyps. At a certain size, these detach and roll away from the parent colony. This is a successful means of propagation which effectively expands the reef laterally.

As in all forms of life, sexual reproduction occurs as well. About a quarter of all species are hermaphrodites (with both male and female polyps), but most have separate sexes. A few, such as the members of the free living mushroom corals of the genera *Fungia*, *Ctenactis*, and *Cycloseris* can change sex as they age, mostly male to female, and in the case of *C. echinata* the corals may later revert back to their initial sex. Most species broadcast both sperm and eggs into the water where fertilization takes place. In some places, and most spectacularly on the Great Barrier Reef, many species do this at the same time, on the same night, in a pattern related to season and full moon, producing such vast quantities of sperm and eggs that thick slicks of gametes and fertilized eggs appear on the surface of the water. Sometimes great drifts of this material get washed up onto the beach if there is an unfavourable wind or current direction. It is assumed that this synchronized spawning behaviour has evolved to swamp predators—providing safety in numbers.

Fertilized eggs become planulae and these can drift in the water column for several days or weeks. Planulae are elliptical and can actively swim for limited distances, although they are dependent on currents for their significant dispersal. They respond to chemicals and light in their environment—chemotaxis and phototaxis. This means that they can choose, to a limited extent, where to descend and attach. Chemotaxis might draw them towards locations with favourable chemicals, such as excretory products from their own species, which indicate that the site is probably suited to their kind. Conversely the planula may use chemotaxis to avoid descending and attaching to an area where the chemical composition may be unfavourable, such as on top of a different species. Phototaxis likewise gives a planula that is ready to settle some ability to determine whether the area of reef over which it finds itself is favourable in terms of amount of light. This requires control of vertical movement to a limited extent, which may permit planulae to remain drifting for a relatively long time while they find a suitable place to settle.

One variant of sexual reproduction employed by some species involves the release of sperm only. The eggs are retained in the parent's body and fertilization occurs internally, with planulae released when they have become more developed. Many of these planulae settle almost immediately, and so tend to grow relatively near their parents.

Coral symbiosis

A remarkable feature of reef corals, and the key to their evolutionary success, is the symbiosis that they form with the single celled algae that grow in the coral tissue itself. These are visible in the sense that their existence in great numbers means a greenish brown colouration is imparted to the entire reefscape. They are dinoflagellates—microscopic algal cells embedded in the tissues of the corals (and in many of the shallow soft corals too). In a sense, when you look at a field of coral you are looking at a field of captive, single celled algae. The algae in this state are called zooxanthellae, and they live in the endoderm (the inner tissue layer) of the coral polyps as well as in any tissue that connects polyps over the surface of a reef-building coral colony. Even though we do not see great stands of plants on a healthy reef, photosynthesis is the basis of this ecosystem too.

This symbiosis is likely to be as ancient as the process of reef-building by corals, and the symbiosis is the reason why corals have become such a successful and dominating aspect of tropical oceans. It permits great efficiency in energy transfer between the sun and the reef constructors. The coral benefits from the products of algal photosynthesis. The algal cells use the carbon dioxide respired by the coral tissue to produce oxygen and carbohydrates in their photosynthesis, which in turn are used by the coral, in a captive cycling process that is clearly very efficient. The densities of these algae can reach millions of cells per square centimetre of coral tissue, and this density may fluctuate over time in response to seasonal variables such as radiance and temperature. Although they are dinoflagellates, they lack flagellae when inside the host. Originally thought to be one species, studies have revealed zooxanthellae to be genetically diverse and, as we shall see later, their study has become increasingly important as the environment around corals alters through climate change.

Zooxanthellae are critically important to the process of limestone deposition too. They are linked to the synthesis of an organic matrix that is secreted by the coral into its sub-epithelial space. The matrix then facilitates deposition of fibre-like crystals of aragonite. Deep-water corals in dark environments lack zooxanthellae and so grow more slowly than their photosynthesizing cousins. The end result in sunlit areas is the deposition of massive amounts of limestone—making up the coral reef.

When corals reproduce, it is crucial that the juvenile obtains its symbiotic algae as quickly as possible. When a polyp divides asexually, or when fragmentation of brittle forms occurs, zooxanthellae are already there, passed across in the parental tissues. But when reproduction is sexual then the zooxanthellae need to be acquired later on via the parent's egg or from free-living algal cells in the water. Surprisingly perhaps, transmission of zooxanthellae into the egg appears to be difficult, and uptake from the water column is more common. The process is poorly understood. Zooxanthellae may aid their own uptake using chemotaxis to locate potential hosts, and when they are in the free-living stage they have been shown to swim towards excretory waste products of corals. When they come into contact with the coral they must be absorbed, but not digested. Once established, the stability of the symbiotic arrangement is actively managed by the coral host, which can encourage or slow algal division and growth as it requires, and each polyp can digest or expel excess algae.

This coral–zooxanthellae relationship is therefore key, but it can be affected markedly by environmental stress. Under difficult conditions the coral will expel its algal cells, and because the coral's own tissue is relatively transparent, the white limestone beneath it will then show through, giving a bleached appearance. Recovery of corals may take place if bleaching is not too severe but, following extreme stress, recovery cannot take place and the entire coral will shortly die. Factors that trigger such stress include extremes of temperature, high or low levels of light, ultraviolet radiation, microbial pollution of various kinds, and many marine pollutants from industrial processes. Bleaching of corals has become extremely significant in relation to elevated sea surface temperatures that are currently affecting the world, and will be discussed later.

Coral food—zooplankton

Photosynthesis provides most of a coral's energy needs, but they are carnivores too. The tentacles of most coral species in all reefs of the world are generally extended only at night. Tentacles are used for feeding, and they are well designed for this. Each one contains many batteries of stinging darts, called nematocysts, with which they capture zooplankton prey. There is differentiation in nematocysts: some seem to be primarily concerned with stinging and injecting toxin, others appear to be more like 'hooks' that latch on to the prey and retain it, and then draw the killed prey into the polyp's central mouth. Although a coral may obtain typically half to three-quarters of all the energy it requires from its symbiotic algae, the polyp can also obtain some additional energy from the zooplankton, though the main benefit of their capture is thought to be provision of additional nutrient compounds that are essential to the needs of the coral. At night, a diver can see how spectacularly rapid the capture of zooplankton may be: if you place a torch beside the coral, zooplankton attracted to the light may touch the tentacle and be immobilized instantly and drawn into the mouth. Some corals are known to be able to sense zooplankton through amino acid residues too.

Those corals that do not contain zooxanthellae have the advantage that they grow down to great depths where there is no light; they can grow in cold water too, even up to polar regions. This shows that symbiosis imposes severe constraints to the range of environments for zooxanthellate corals. The azooxanthellate forms are usually minor components of tropical reefs as they grow slowly and are generally outcompeted in sunlit areas, but they are common where it is dark, such as on deep and steep slopes, on the undersides of overhangs and in dimly lit caves. Many are solitary and small but some species do form large, dense aggregations.

Even though these corals depend entirely on zooplankton capture by their tentacles for their nutrition, shallow dwelling forms commonly retract their tentacles during the day in the same way that their symbiotic cousins do, and this may be both because of the threat from grazing, and because most plankton only emerge from the substrate at night anyway. The number of species of azooxanthellate corals is as great as those with symbionts, and they are a generally interesting but minor component of tropical coral reefs.

Coral wars

In normal conditions, there is intense competition for space on the reef between the many species of corals, and between corals and other organisms. While it is easy to see activity, aggression, feeding, and flight amongst the fish, for example, corals and soft corals look far more static. But that is only because of our human perception of time. They too compete for space, using several different mechanisms to aid them in their competition for survival. Most of this activity happens at night. You can see its effect by observing apparently bare zones or dead 'halos' around many or even most corals. More spectacularly you may see a distortion or hole in a table coral because lying beneath it, or to one side of it, is a less conspicuous brain coral, for example, that is preventing the table's growth over it. Soft corals also are commonly seen with narrow bare bands around them, a result of the toxins that they secrete. These and other methods both prevent other species from overgrowing them and create bare space around them into which they can expand.

This competition for space may take at least a day, but more commonly two or three weeks, depending on the mechanism that the coral employs (Figure 7). The first mechanism involves the dominant species exuding digestive mesenterial filaments onto a neighbour. This is a short-range but very rapid mechanism, which will be triggered within one or two hours if

7. Coral competition. Top: a polyp of *Lobophyllia* has extruded its digestive mesenterial filaments over an adjacent faviid colony. By morning, most of the coral on the left will be bare skeleton. Bottom: sweeper tentacles of the brain coral have prevented two different species of table coral from overgrowing it. The normal tentacles of the brain coral, and these valley widths on its surface, are less than 1 cm, but the sweepers which are deployed at night will be 10–20 times longer

two corals are placed adjacent to each other—in an aquarium for example. It mostly takes place at night under natural conditions, and there is usually a consistently dominant and subordinate species in any one pair of species. The dominant coral digests the subordinate species, leaving only bare skeleton by morning. A second longer range mechanism involves development and deployment of sweeper tentacles, ten or even 50 times longer than normal tentacles, that are heavily loaded with stinging cells called nematocysts that inject toxin. The tentacles are swept over and onto a nearby subordinate coral, stinging and killing it. These sweepers take up to a couple of weeks to develop following detection of a neighbour, but their long reach is critical. Developing these tentacles obviously requires a considerable investment of energy, so the fact that they go to this trouble shows its importance in maintaining space. A variation on sweeper tentacles is the development of long sweeper polyps, which has more or less the same effect.

Some families of corals may be very aggressive, while others seem to be subordinate to just about every other species. This being the case, in order to survive, they must have other, superior, attributes such as faster growth or greater reproductive rate.

All this leads to considerable spatial movement on a reef, although during the course of any one dive an observer cannot possibly be aware of it. However, in measurements over two years, I have shown that although the percentage cover of corals, bare space, and so on may remain fairly consistent, beneath any one point there is repeated change from coral to bare space, and back to coral, perhaps of another species, at a rate of perhaps two or even three changes in a two-year period. Movement on the reef may take place in slow motion in human terms, but nonetheless reef surface occupancy is very dynamic.

The soft corals

Already mentioned have been the soft corals, a widespread group of animals related to the corals and intermixed with them on most areas of reef. These are octocorals, with each polyp being separated into eight segments, each with eight feathery tentacles—unlike the six or multiples of six tentacles seen in stony corals. As their name suggests, their skeleton is organic and flexible. These corals are especially abundant in the Caribbean (Figure 8). Amongst this group are the sea fans, which are colonial animals that hold their polyps high up into the water column to

8. An assemblage of many soft corals on a Caribbean reef at about 10 metres depth. British Virgin Islands

feed on passing plankton. Some soft corals also contain symbiotic algae, but not generally to the extent of the stony corals, being more dependent on plankton capture. They occupy significant amounts of space on a reef and are ecologically very important. However, they do not help in reef development and their soft colony structures disintegrate after death.

Other calcareous coral relatives

A variety of other calcareous animals may be abundant as well. The organ pipe coral *Tubipora musica* is an anthozoan of a different order, with a hard bright red skeleton made up of small vertical tubes, each housing a grey polyp. The blue coral *Heliopora coerulea* is of yet another anthozoan order, and has a calcareous, bright blue skeleton, though these are covered with brown tissue when alive. This species is most common in shallow water, and both of these species are found only in the Indo-Pacific. Then there are the more distantly related fire corals of the genus *Millepora*, which are in the different coelenterate class Hydrozoa. These are common to both Caribbean and Indo-Pacific reefs. These earn their name because their powerful stinging cells can penetrate thin human skin causing a burning sensation, and although fire coral stings are not usually dangerous, some people may show a severe allergic response to them.

Sponges

On many reefs, sponges may be particularly important. Sponges are extremely ancient multicellular animals, whose ancestors extend at least as far back as the Cambrian Period over 500 million years ago. They are enormously varied in shape, even within one species. Some form huge vases, others are encrusting or consist of solitary pipes. Some ecologically important sponges burrow into limestone and are important eroders of the reef. They are important for several reasons. They occupy significant amounts of substrate especially in the Caribbean, and with over 5,000 species so far identified worldwide. With many more species awaiting identification, their diversity is much greater than that of the corals, which perhaps reflects the longer time over which they have existed. Sponges feed by pumping water through their bodies to extract particulate matter including bacteria down to picoplankton size (0.2–2 μm), and there is recent evidence that sponges even have the capacity to capture viruses. Their ability to clear water of particles is prodigious: for instance, in the Caribbean a range of sponge species were found to remove two-thirds to nearly all the particulate matter from the water column above them, and sponge communities at 25 to 40 metres deep have been estimated to filter the entire water column above them every day. This capture of plankton is a key link between the water column and the benthic life (life on the sea bed). This 'bentho-pelagic coupling' provides an important route through which particulate carbon and nitrogen are channelled between water and the reef.

Other sponges are important bioeroders, causing disintegration and erosion of the reef by boring into reef and coral limestone through secretion of acids, which produces large quantities of fine sand.

Calcareous algae

One crucial group of reef constructors is stony algae. Amongst the red algae are found possibly the most important species: stony species that thrive on reef crests to create strongly wave resistant ridges (Figure 9). These literally permit the coral reef to persist in the face of storms and strong waves. The reef crest lies at the seaward edge of the reef flat and marks the point between the reef edge and the slope that plunges steeply. Very few coral species can grow in this most wave-exposed part of the reef. These calcareous reds actually require strong water movement and high aeration, and some calculations show they are as strong as concrete. They form a ridge that grows along the edge of the reef flat, from which huge spurs extend several metres: this is known as the spur and groove system. The spurs dip gradually into deeper water, and

9. An example of a spur and groove system at the edge of the reef flat, marking the point where the reef slope plunges downwards. The ridge and spurs are constructed from huge growths of calcareous red algae, and form a most effective barrier to destructive wave energy

have steep sides, usually with a scoured appearance, and fade out at a depth where average wave turbulence is greatly diminished. The mechanism regulating the spacing of the spurs is intriguing and thought to be some function of the wave energy received at that point, such that the greater the energy from the wave, the larger and more widely spaced the spurs. Whatever their size, the structures greatly reduce the impact of breaking waves in a self-regulating arrangement, and the swash-back of each wave in each groove collides with the next oncoming wave in a way that is both spectacular and yet dissipates much of the wave's energy. This feature is key to the existence of reefs in areas of high wave energy—i.e. most oceanic reefs.

Chapter 4
The resulting structure—a reef

Reef profiles are remarkably consistent, which should not be surprising when it is remembered that they are essentially biological constructions responding to environmental factors.

Reef flats

By far the largest visible expanse of reef when seen from land is the reef flat. Its area usually dwarfs that of the reef slope that descends from it into deeper water and, for a couple of hundred years, the early naturalists focused on this component more or less exclusively, through necessity. Photographs exist of naturalists dressed in plus-fours, jacket, collar, and tie in the tropical sun, perhaps with an umbrella and a walking stick, posing on reef flats. This to them was the reef. Figure 10 illustrates typical cross-sections of two typical reef flats although the extent of the horizontal portion is made narrow to fit the page: reef flats may be over a kilometre wide. Most early naturalists recognized the existence of a profusion of life deeper down, but it remained inaccessible—the *mare incognitum*.

There are reasons for the large extent of most reef flats. Oceanic tides usually have a range of a metre or more, so that at high tide every reef is completely submerged so that you can take a small

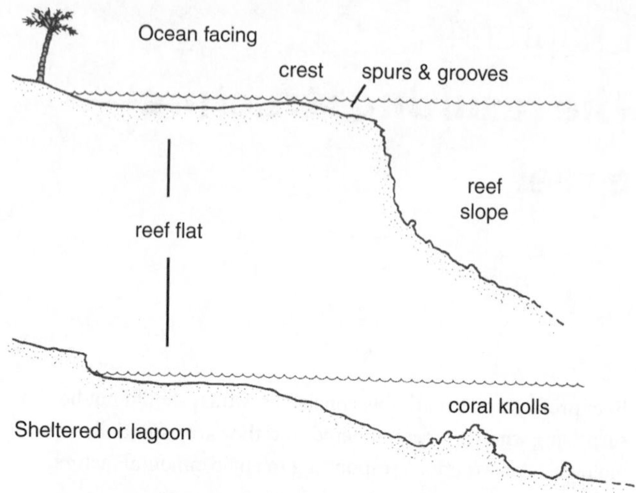

10. Diagram of the cross section of a coral reef. Top: exposed or ocean-facing reef, and bottom: a more sheltered reef. These are typical profiles and most reefs show little variation from this general scheme

boat across it or snorkel across it. However, at low tide the reef flat will dry out. This is a uniform characteristic seen all around the world and one that led to much speculation in the 1800s about why such structures were so consistent.

They are consistent because reefs are built by marine life and related processes, and construction can occur only underwater, not in air. This inevitably leads to structures that grow up to low sea level. Being so shallow, this area regularly experiences environmental extremes. The high temperature to which the reef is exposed during low tide, severe dilution of salinity by rain during the monsoon at low tide, alternating with dangerously raised salinity under the tropical sun when no rain occurs, not to mention severe desiccation from sun and wind as well as the effect of damaging ultraviolet light, are all enough to ensure that only a small sub-set of reef life has become adapted to live there. Biologically, therefore, reef flats are relatively depauperate. They are an environmentally stressful 'outpost' of the heart of a reef. This factor greatly limited early biologists, because to understand a coral reef from its reef flat alone is akin to trying to examine the Amazon rainforest only from the roadside; you simply cannot see the diversity beyond. Some of the species that have adapted to its extremes may be very abundant, but their diversity is low and many reef flats are relatively devoid of life. It may be an interesting place to research physiology of corals under stress, but this place is not overall representative of a coral reef in ecological terms.

The types of life that we see on the shallow reef flats are either outliers from the main populations deeper down or, in some cases, they are forms that are so well adapted to this location now that they are very rarely found elsewhere, perhaps because they are easily outcompeted everywhere else. Informative examples of outliers from the main populations are the massive shaped colonies of *Porites*, and occasionally genera from the Faviid family, which includes the brain and star corals, which can be seen growing on the reef flat in characteristic 'micro-atoll' form. A micro-atoll is a single colony that has grown into a ring with living sides but with a dead and eroding top. They cannot grow upwards anymore, only outwards, expanding their girth, and providing clues in connection with the study of sea levels over the life of that coral; fossil micro-atolls have also been found and are used to help in reconstruction of ancient sea levels.

Reef flats therefore develop because coral reefs grow to the low water mark and then expand outwards. One well-developed example is seen around the island in the foreground of Figure 3; some reef flats may extend a kilometre from the shore before the reef descends into deeper water. Although these large horizontal expanses are commonly amongst the largest components in terms of area, for corals they are mostly rather marginal habitats.

At the seaward edge of the reef flat, and commonly a little more elevated above sea level than most of the reef flat, lies the reef crest. This is usually marked, and made, by massive ridges of pink, limestone-depositing algae—the coralline algae described and illustrated in the previous chapter. Here there may be extreme turbulence, to the extent that almost nothing else can live there, because the waves break upon it. These structures absorb perhaps 90 per cent of the ocean's energy that otherwise would strike the shoreline of the island.

The reef slope

To seaward of the reef crests, the reef slopes more steeply downwards to various depths. Reef corals continue to grow deeper and deeper until light becomes too dim for continued photosynthesis, after which ancient reef may continue deeper still, or sand covers the surface. Together with reef flats, this broad structure indicates an actively growing reef. The pattern seen is complicated by changes and shifts of sea level over geological time, so the basic pattern commonly will have superimposed upon it evidence of episodic growth and erosion, which complicates the picture.

The environmental conditions on the reef slope are ideal for most kinds of corals and other reef life including the soft corals and sponges, and this is where most species may be found. Peak diversity may occur anywhere between 5 to 20 metres deep, depending on the area and overall conditions, particularly overall water clarity. Here, sunlight is still bright enough for photosynthesis but not damagingly intense, and wave energy is not destructively high but water movement is still sufficient to keep sediment from settling. Put simply, most reef species want to live here, where environmental factors are moderate, and in order to do so they must show a competitive edge in one or more aspects of their lifecycle—competitive ability, growth rate, or reproduction perhaps. Coral reef slopes support the richest biodiversity and most varied of macro fauna in the seas. It is crowded!

Deeper still, at the foot of the reef slope, light begins to dim, and in order to trap light, leafy forms of corals increase and dominate. Wave energy fades to almost nothing, meaning thinner, leafier, and more fragile corals can survive. Their angles are usually such that they trap plenty of light, but their leafy sheets still slope steeply enough to shed sediment, and in many parts of the world, the Caribbean notably, the deep reef slopes will be dominated by 'shingles' of corals like the tiles on a sloping roof. Eventually, at depths which may be greater than 80 metres or more, all photosynthetic forms disappear, and the true reef can be said to have terminated, sometimes on to sand, or perhaps onto rocky surfaces (which itself may be ancient reef), being dominated by forms which do not require sunlight.

Cryptic life

A large proportion of reef life is cryptic, that is, occupying crevices or making its own crevices by tunnelling into the soft rock. When it is remembered that about two-thirds or more of the fishes that swarm over a coral reef are carnivorous, many with very wide food preferences, the desire to hide becomes easy to understand.

A hidden or cryptic way of life has an obvious drawback, however, namely reliance on food that comes towards it. Hence, many cryptic species are filter feeders, trapping drifting food, and many prey on those that have done this work for them. Many worms, crustaceans, and molluscs do this, and sponges are exceptionally important filter feeders too. Tiny predators, including crustaceans and molluscs, prowl the maze of subterranean caverns and passages, while other species carry protective shells around them—molluscs and crustaceans being prime examples. For every development, though, counter-measures have been developed: some crustaceans crack or peel open the shells that molluscs create to protect themselves, while some predatory molluscs drill holes in others to get to the flesh beneath. A continual arms race is taking place on the reef, in which one species' defences are countered by another species breaching the defence; and each new kind of attack is countered by new kinds of defence. The 'Red Queen' analogy is very pertinent for reef life and is seen in many guises. This term is derived from *Alice in Wonderland*'s Red Queen who had to run continually just to stay in the same place. The species that exist today have survived as a result of successful strategies in this arms race. This is self-evident in one sense: if the species cannot evolve continuously, or keep itself safe from predation and elimination, it becomes extinct—the fate of 99 per cent of all species that have ever lived. And in the most vibrant part of the reef, where environmental factors are relatively benign, competition is intense, and one defence has been to remain as unseen as possible.

Rugosity, a key feature

A consequence of the myriad growth and building behaviours of corals is a reef which, like a forest, provides a three-dimensional habitat: a large and varied space in which to live. This is thought to be one reason why reefs support such a high diversity of different forms. Quite unlike, say, a flat sheet of limestone, irregular coral growth and the colonies themselves provide considerable volume and countless niches. Most striking in this respect is that most architectural of all coral genera, the *Acropora*. While several coral genera contain some species which develop branches, none match the *Acropora* in their ability to form thickets, miniature forests, and table-shaped colonies, providing much space for hiding, ambushing, or escaping. Coincidentally, perhaps, the genus *Acropora* is the most speciose of coral genera with more than a hundred species described. So important is the structure that they and all corals provide that methods have been developed to try to quantify it and relate the three-dimensional 'rugosity' to various characteristics of reef biology. A diver can see obvious indications of the importance of this by the numerous schools of small fish living in branching coral colonies; they feed on plankton close to the coral but dart back into the shelter of the corals' branches when anything much bigger than themselves approaches. From these small fishes to the largest grouper, many species rest or wait in ambush under tables of *Acropora*, whose wide range of niches provides shelter to both hunter and hunted alike.

The branching structure comes at a price for colonies that display it, the most obvious being loss of physical strength. Because of this, most species of branching corals are found where wave energy is not at its strongest, namely several metres deep, on reef slopes and in lagoonal areas. But there are very stout species of *Acropora* too, in both Caribbean and Indo-Pacific regions. In the Caribbean, the Elkhorn coral *Acropora palmata*, which is the largest coral of all in many respects, can dominate the shallowest areas all around islands. Or at least, it used to do so; it has recently suffered devastating diseases that have all but eliminated it—with severe consequences—but it still exists, and is said to be recovering in one or two places. Slightly smaller but equally wave resistant relatives are found in the Indo-Pacific, including *Acropora palifera*, which similarly occupies the shallowest and most turbulent water, though these too have suffered recently, this time because of warming water.

The massive shaped corals, by which we mean rounded, boulder shaped forms (even where they are not particularly large) also add three-dimensional structure. Deeper down are leafy corals, extending outwards to capture light. These usually have small and widely spaced polyps with poor plankton capture ability, relying heavily on light capture for their zooxanthellae. Some of the deepest leafy forms appear to have almost no polyps at all and are the most heavily reliant on their photosynthetic surfaces, but probably all depend to some extent also on organic debris that settles on to their surfaces.

Caves, hollows, and indentations in the reef add an additional level of complexity. Whether taking the form of a simple small overhang or a deeply penetrating cavern, their fauna is markedly different to that found just outside it. Horizontal runs of caves occur on many steeply sloping reefs, created in the geological past when sea level was tens of metres below its present level, when that part of the reef was pounded and eroded by waves. Or, many caves and fissures may be solution features caused at the same time, namely when sea level was a long way below where it is at present, when copious rainwater etched tunnels that later became drowned when sea level rose again. Whatever the cause, caves show a rich abundance of rather different dominant forms of life, including those usually seen only very much deeper, where they are too deep for most divers to see.

Caves show a marked zonation as well. Around entrances will be the kinds of corals and soft corals that are typical for that depth on the reef, but, further in, and as light diminishes rapidly, photosynthetic forms rapidly change to red algae in the case of the plants, from zooxanthellate corals to those without symbiotic algae, and to hydroids and sea fans, and maybe some black corals too—a different group of colonial animals with a skeleton much sought after for polishing into jewellery. The commonest group of attached animals are usually hydroids, most species being feather-like structures forming colonies of small animals that pack powerful stings easily capable of penetrating human skin. Lining the walls of the caves there also may be numerous species of sponges. Amongst and around these sessile animals, and indeed within many of the cavities in sponges themselves, are numerous polychaetes, crustaceans, and other species, specialized for cave dwelling life and different from their more numerous relatives outside the caves.

Being a dark environment, the fauna attached to the walls of caves is entirely dependent on plankton, both alive and dead, or on the waste products of those that do feed on it. The origin of the plankton may be quite obscure. Most presumably are pumped in by water movement from the outside, and so a cave which only has one aperture may be relatively barren. But many caves have heavily fissured walls and roofs, too small to notice visually, but measurable by instrumentation that analyses water flow. Many reefs turn out to be remarkably porous, permitting considerable flow of water through them. Even leaving aside small chimneys through which light may be seen, many caves experience the passage of thousands of litres of water per day through numerous small channels, pores, and crevices, and to a considerable extent it is this permeability of the reef rock which determines the amount of life that can live in a cave. This can be demonstrated easily, if unscientifically, by a diver inside the cave purging his or her regulator. The air disappears into the roof, and a dive buddy outside the cave might see, after a pause of maybe a few seconds or minutes, streams of air bubbles emerging from the apparently unbroken reef surface several metres away from the cave mouth: and where air can pass, so can water containing plankton.

Caves may not be permanent, and indeed nothing much on the reef is permanent over the longer term. Growth of calcareous organisms and the pumping of sediments into a cave act to close it, even if only very slowly. Alternatively, and depending on location and depth, wave action may enlarge it. A cave roof may collapse from the weight of coral above it, and, while not a common sight, it is possible to see places where an avalanche of reef has crashed down-slope, scouring a substantial track behind it which is now filled with sand. I have seen an extreme example of this where the typical reef formation of reef flat, crests, and slope was completely changed by an earthquake that occurred 150 years ago (described by Charles Darwin, whose account alerted me to take a look at that spot). That earthquake took with it not just a large chunk of reef but a portion of an island above it as well, the process being noticeable to the people who lived on that particular atoll because of the resulting loss of part of a coconut plantation.

Caves, fissures, and tunnels occur because coral grows vigorously in the light, and when it grows outwards it can overgrow its own foundation, so to speak. Judging by human observation, this happens mostly in slow motion, like most things about corals, but the reef has been around for millions of years and such overhangs, caves, and vertical slopes all add different habitats for different groups of species.

Sand creation

Probably the largest expanse of any coral reef is its back reef area, composed mainly of rubble and sand, where few corals grow. This is an integral part of every coral reef, vital ecologically and interconnecting with the reef, both in physical and biological senses. The sand is produced by the reef life and it is truly a part of it.

From one perspective, the principal role of corals is to grow, die, and become broken down into ever-smaller lumps and particles, from which are built up both the coral reef matrix and coral islands themselves. From the moment that they begin to grow, corals are subjected to both biological and physical factors that act to reduce them to sand.

Physical erosion is straightforward. Storms and wave action readily break up colonies, especially branching and leafy ones. Fragments produced may initially be fairly large: whole branches, fronds of leafy colonies, and fist or finger sized nodules. These, when rolled around by succeeding wave action, gradually become reduced in size, producing fragments that eventually become turned into sand and then silt, which accumulates behind the reef and in its lagoons. It also accumulates in the numerous small spaces on the reef itself, where it later becomes consolidated into more durable rock. Substrate covered with this rubble is very inhospitable to most forms of life that require a solid and stable base because its movement in storms has a similar effect to that of liquid sandpaper, producing a surface that is very hostile to most attached forms of life.

The biological attack on the skeletons that produce sand is more complex and varied, and generally results in the creation of much finer particles and silt. Parrotfish swarming in large schools over the reef graze with beak-like mouths, scraping the surface, taking around 1 millimetre-deep bites from the substrate each time. Many large boulder corals may be seen covered with these scrape marks, standing out white against the brown background of healthy coral tissue. Parrotfish are responsible for the production of much of the sand on reefs. Their digestive systems are such that they require sand to aid digestion, and it is common to see a stream of fine white sand being defecated from a parrotfish as it swims.

Far less visible is bioerosion by cryptic forms of life, usually very small animals, and by some plants as well. Although largely unseen, the cryptic reef fauna are estimated to have a biomass as great as that of the more visible fauna on the surface, and it has a much greater diversity of species. Small, bioeroding organisms tunnel, bore, and chisel into the relatively soft limestone rock. Many, such as bivalve molluscs, use their shells to scrape out a tunnel in which they live. Other organisms secrete mild acid to dissolve the limestone. A particularly important bioeroding group is the sponges. These have no mechanical or moving parts and so rely entirely on dissolving the rock. Some develop extensive but thin sheets of tissue covering the surface, distinguished by many irregularly spaced, small volcano-like structures, which are the exhalant pores of the sponge, whose main biomass lies beneath in an increasingly hollowed out coral. In the Caribbean, some large sponges of the genus *Cliona* (Figure 11) can erode many kilograms of substrate in a year from coral colonies that they have colonized. Net bioerosion from sponges has become a dominant factor in many reefs where corals have been killed for whatever reason, but most corals on a healthy reef are not colonized in this way.

Plants also bioerode. Filamentous green algae, especially, work their way in, often using the minute, naturally existing pores in any coral colony. Being photosynthetic, they need to remain in the top few millimetres where they still receive some illumination.

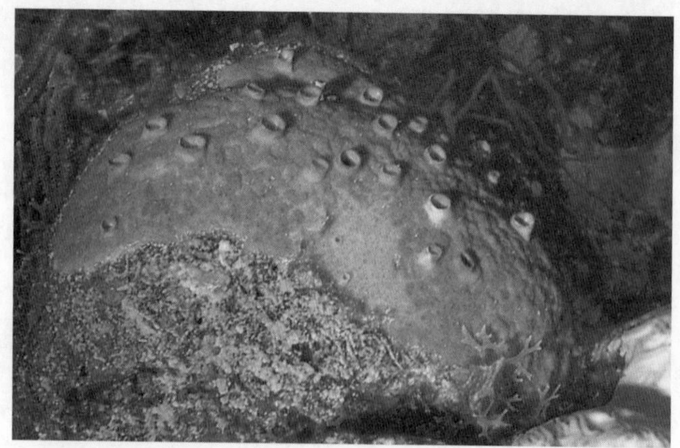

11. A coral colony being destroyed by the sponge *Cliona*. The sponge, bright red, is encrusting only on the surface; most of the sponge is inside the increasingly hollowed-out coral limestone

The translucent quality of corals' aragonite limestone means that light can penetrate sufficiently through both a thin layer of coral tissue and a couple of millimetres of limestone for sufficient photosynthesis to continue. A section through many massive corals will show a thin green band that is indicative of the presence of such algae. One of the main groups is a genus called *Ostreobium* and, because this group lives so near the skeleton's surface, it has been suggested that they may even contribute in some way to the nutrition and gas exchange of the coral in a way similar to the symbiotic zooxanthellae.

Overall rates of erosion of reef and corals may be close to the highest levels of calcification or growth that have been recorded. Growth and erosion are usually finely balanced and the usual situation on a healthy reef is one of net growth, if only just, which is why coral reefs tend to reach the surface in the first place. But any interruptions to growth, or increase in erosion, can easily result in an imbalance, leading to destruction of the reef.

Another factor is important. Polyps cover the surface of a living coral, and polyps are carnivorous. Any unfortunate larva of a tunnelling or boring organism trying to settle on a live coral is consumed, so that the live surface of any coral acts as a most effective barrier to penetration of its limestone skeleton. Mortality of the coral tissue, however, immediately exposes the skeleton to attack, and leads to greater sand and sediment production. It is not only a psychological gloom that makes a diver think that a dead reef has poor water visibility, but also the increased quantity of very fine silt suspended in it that has originated from such erosion.

Many other groups of organisms produce sand, some in significant quantities. Foraminifera, or forams, a group of protists, have shells that usually are limestone that form sand grains less than a millimetre across to perhaps a centimetre, depending on the species. Some of these protists contain symbiotic photosynthetic algae as well. The forams of greatest importance in the context of sand are the benthic species, and may be so abundant that they can comprise most of the sand over extensive areas. Colonial species of forams create encrusting sheets and play an important role in helping bind together the sand and rubble beneath them: they can create huge sand banks.

Other groups of animals contribute to the sand in relatively minor ways and may be distinguishable by having colour, sometimes vivid, in the otherwise pure white sand grains. Mollusc shells, echinoderm tests and spines, and skeletons of the blue coral *Heliopora* and the red organ pipe coral *Tubipora* all become part of the sand, adding flecks of reds, browns, and blue.

The sand produced plays positive, constructive roles on the reef, a critical aspect commonly overlooked but which is vital to the growth of the reef. Sand and silt settle into crevices and pores, sometimes forced in deeply by wave action. These particles consolidate and slowly turn the reefs into consolidated limestone rock that is more durable than the coral colonies from which the particles were derived. A cross-section of an exposed fossil reef will reveal as much or more of this amorphous looking solid limestone matrix than do recognizable coral colonies. How these particles consolidate or become fused into solid rock may have something to do with microbially mediated changes in pH on the surface of the grains, but the mechanism is largely unknown. However, the result, which can be seen in an exposed fossil reef, for example, is a structure of numerous dead coral colonies of many species surrounded by and embedded in a durable rock matrix. It is this, principally, that creates the geological structure of a reef.

Reef sand is also made directly by several kinds of plants from all three major categories of large algae—the greens, reds, and browns. The most important of these is the green *Halimeda*, an unmistakable plant consisting of chains of small limestone discs covered with living, green tissue, each disc being round or kidney shaped, and a few millimetres to a centimeter wide or more depending on the species. Each disc is mostly limestone, and each frond may produce a new disk every day under ideal conditions, with one plant developing dozens of chains. Connections between discs are fragile and break easily; the tissue dies and the discs become a readily recognized part of the sand. Their production of sand is prodigious, such that many Caribbean beaches are largely made up from this and, in the Pacific too, enormous mounds have been discovered in depths of several tens of metres. Vertical underwater surfaces may be covered with *Halimeda* plants from which stream billions of discs daily, destined to become a dominant part of the sand and an important part of the whole reef complex.

A few brown algae also secrete limestone within their fronds though nowhere near to the extent of *Halimeda*. The genus *Padina* is typical of these, though the quantity of limestone it produces is much smaller and very fine. Red algae likewise include several genera with similar limestone secreting properties, but the most important algae in this taxonomic group are those very stony genera that develop the algal ridges and spurs described in an earlier section.

Sand life

The quantity of sand produced on a reef generally exceeds several kilograms per metre square per year. While much becomes entrained in the reef, much more drifts into back-reef areas to create expanses larger than the reef itself. Important sand-sorting processes happen continuously, driven by wave or current energy whereby sand is continually graded from coarse to fine. Immediately behind the reef is generally the rubble and coarse sand, leading to ever finer sand and finally possibly to calcareous muds. This sorting occurs primarily because finer particles remain suspended in water for longer and settle out more slowly, so are carried further in the water flow, which is driven by waves breaking on the seaward edge of the reef. It is commonplace for reefs to grow into prevailing wave energy, so that there is almost continuous movement of water from the reef front or reef slope, across the reef flat, to areas behind. The sand may become piled up to form a beach around an island, sustaining that island or even creating a new one. Currents across the top of a reef may commonly flow at more than one metre per second, more than enough to carry suspended sediments considerable distances, but when deeper water is reached the velocity immediately slows, allowing grains to progressively settle out according to size.

This sand is far from lifeless, and different sized sand particles support different assemblages of species. The first point of note is that the sand underwater very often appears brown, grey, even purple, and mottled—not white as when it is piled up to form a beach. The colour is caused by expanses of filamentous algae and microscopic forms such as diatoms growing over the surface, photosynthesizing, and providing food for numerous other organisms. Many of these also fix nitrogen. Commonly mixed with these are enormously productive patches of larger algae and seagrasses. All of these add organic matter to the substrate, which is consumed by numerous larger forms of life. Many species contribute to the sand itself, aside from the forams noted above. The most numerous animals are small, for instance the so-called micro-molluscs, which are the size of sand grains themselves. Much more conspicuous users of sand may be sea cucumbers of the echinoderm Holothuriidae family, which feed on the organic matter by ingesting vast quantities of sand, digesting its organic content and excreting the rest. In those increasingly uncommon places where they are not collected for Asian food products, their numbers might exceed two or three in every square metre; they are a crucial component for 'cleaning' the sand. Where they are heavily collected, organic matter can build up to levels which lead to localized anoxia.

Numerous molluscs, including many predatory species, burrow within the sand, some surfacing only at night with their siphons extended above the surface during the day. Many flattened forms of echinoderm called sand dollars and heart urchins also burrow. But some of the most conspicuous forms of these various groups are worms, notably polychaetes, many in tubes with one end exposed to the surface. Several fish burrow into sand. Striking examples are the garden eels, shy creatures forming a sub-family within the conger eel Congridae family, whose colonies cover thousands of square metres, and which withdraw tail first into their burrows at the slightest disturbance, including the approach of a diver. Burrowing shrimp may be common and conspicuous sand dwellers, as are rays, partially buried beneath the top layer, with just a pair of eyes visible.

Burrowing by such species is enormously important because it ensures that sub-surface sand and silt is regularly mixed with surface sand, bringing it up to the top where it is exposed to oxygenated water in a process that prevents an anoxic layer developing below the surface. Experiments in which black sand is spread over areas of white coral sand have shown that a thorough mixing of surface with sub-surface material takes place within just a few days.

Many, if not most, families seen on a reef have relatives adapted to living on and in the adjacent sand. This area remains much less studied than the reef itself, which is unfortunate as it is a very important component of the wider reef ecosystem. The sand may build up an island, something of huge commercial and tourist importance, not to mention in providing a home for people, and these sometimes huge sandy expanses provide the principal feeding area for several species of fish used by local communities. Such areas should be viewed as being an integral part of the reef. Many fish of commercial importance migrate back and forth between sand and reef, some on a diurnal basis and others on a seasonal basis, to forage for food, forming an ecological link between the reef and its adjacent sandy areas.

Chapter 5
Microbial and planktonic engines of the reef

We know that the symbiotic algae are a crucial source of fuel for the reef, via, for example, the corals, and there is a tight recycling of nutrients between the coral animal and its symbiotic algae. There is little 'leakage' out of the coral–zooxanthellae symbiosis. But transmission of this energy to the rest of the system is not explained by simple feeding. How does it enter the ecosystem to support the huge abundance and diversity seen there? Important engines of coral reefs, indeed of any ecosystem, are the mechanisms by which organic material produced by photosynthesis is transferred along its various food chains, from plants to the top carnivores in the system. This chapter describes the usually unseen engine that transfers food and energy from the corals to the other parts of the ecosystem.

Particles and plankton

The major mode of feeding by a large proportion of reef animals is filter feeding—extracting particles from the water. The zooxanthellate corals do this too—which is one main purpose of their tentacles with stinging cells—even though corals may obtain most of their energy from their photosynthetic symbionts. Soft corals also do this, but they need a greater proportion of plankton to supply their energy needs. Sponges, and many other animals such as bivalve molluscs, featherstars, countless worms, and basket stars amongst many others, feed heavily on the plankton, either by pumping water through their bodies and sieving out useful material or by draping tentacles in the water currents to trap the particles as they pass. Some species are able to discriminate not only the size of plankton but its material too—a particle of fine sand has no nutrient value, of course, and needs to be rejected, whereas a zooplankter of the same size is retained. Some of the smallest fish are known as plankton pickers; these literally see and pick individual zooplankton suspended in the water, while larger sea creatures up to and including manta rays and whale sharks simply scoop in huge volumes of water and filter it. Large zooplankton feed on smaller zooplankton, and smaller zooplankton feed on plant plankton. Their primary food source may be the plant plankton—the phytoplankton, but much comes from microbial processes of breakdown.

Microbes and viruses. The smallest of all plankton are the truly microbial forms whose numbers are staggeringly high: even in clear oceanic water there may be a billion in every litre. Many are free living while many others adhere to particles. The bacteria, archaea, viruses, and protists that comprise this group are descended from the first forms of life, and have existed throughout the time that life has been recorded on earth, so it is no surprise that in terms of diversity, biomass, and metabolic activity they exceed all others. These have a key function on the reef: they break down organic matter and recycle it, mineralizing some nutrients and growing on the material to become food themselves. The doubling time of many kinds of cells is prodigious and ten doublings per day have been recorded. They can show very rapid daily fluctuations in density as they respond to new food sources, being most numerous during daylight when they respond rapidly to photosynthetic products of algae and mucus from corals. Their total productivity on a reef floor may be half a gram of carbon per metre square per day.

These microbial communities, especially bacteria, are key to the recycling of carbon and organic compounds on reefs, and this is known as the 'microbial loop': the invisible aspect to how a reef functions. Key nutrients such as phosphorus in the water over a reef can be turned over in minutes, and even complex organic materials are recycled in hours or days. For many years it was considered paradoxical that coral reefs could exist as oases of high productivity in the nutrient-poor vastness of the open ocean, but it is clear now that one reason why this is possible can be attributed to the microbial loop.

Direct consumption of vast numbers of microbes that have grown fat on waste and dead organic matter is very important to many species. Bacteria commonly clump around waste matter forming larger particles, and these bacterial clumps become the food source for many filter feeders and for thousands of tiny scavengers on the sea bed. Many of these are single celled animals themselves, but many others are minute, multicellular forms. Consumers of these bacterial clumps also include the zooplankton, some living in the water column and others living between grains of sand. Most of this minute fauna can grow rapidly in response to the bacterial food source, and even large scavengers such as sea cucumbers utilize the material, as they feed by ingesting large quantities of sand.

Marine viruses cannot be ignored, but these are still a rather enigmatic component on reefs. They occur in densities of up to 10^8 in each millilitre of seawater, may be more abundant still by a couple of orders of magnitude on the seabed, and probably infect all organisms. It is now known that the filter feeding apparatus in some sponges is capable of extracting some viruses from the water directly. It is presumed that these viruses can greatly affect many ecological processes, just as they do on land, including regulating populations and communities.

Slightly larger than the bacteria are the phytoplankton, mostly single plant cells. These are relatively scarce in very clear water but occur in high densities in the water column in reefs nearer land, where perhaps there is nutrient runoff acting as fertilizer. Diatoms and dinoflagellates are important components. The dinoflagellates are especially known for their symbiosis with corals, as we have seen, but many are entirely planktonic and many more live on surfaces of rock, seaweeds, and seagrasses. They are a complex group. Some are motile and entirely predatory, feeding on other protozoans, and use their two flagellae to swim. About half of the known species photosynthesize—some of these are the zooxanthellae of corals. Sometimes the free-living forms undergo blooms, commonly called red tides, which may produce neurotoxins that result in massive fish kills and produce ciguatera poisoning in humans who eat the contaminated fish.

The densities and diversity of dinoflagellates and especially diatoms on the surface of sand are immense, with several hundred species in the top centimetre or two of a sandy patch behind a reef, living at densities which, in one study on the Great Barrier Reef, revealed two to three million diatom cells in every millilitre of sediment, providing up to a gram of chlorophyll in every cubic metre, and a significant total biomass. These microscopic algae may contribute a quarter to a third of the total primary production on a coral reef.

Protists, which used to be called single-celled animals or proto-zooplankton, both terms which describe them well, are the next step up the microscopic food chain. These are abundant and several are major grazers of much of the phytoplankton, including the benthic forms. Grazing of phytoplankton by vast numbers of protists is a major link in the coral reef's trophic web, responsible for much of the carbon transfer to larger organisms. They are diverse, occurring in both seawater and sediments. Ciliates are important too, and many of these also contain symbiotic dinoflagellates, so as well as consuming smaller plankton, some also increase primary productivity as well. Finally, foraminifera are a large group of amoeboid organisms best known for existing in huge numbers on the seabed but which also have many planktonic forms as well. These along with radiolarians, can develop very elaborate external skeletons, and skeletons of both groups have been very useful in tracking past climatic changes. With these forms, much of the energy and carbon moves up the numerous food chains of coral reefs.

Temporary zooplankton—larvae. Larvae of larger animals form part of the plankton too. The purpose of the larval stage of a sessile animal is obvious: it is the dispersal phase. The largest larvae are usually those of sessile species such as corals. Many can swim to a certain degree but more important perhaps is their ability to control their vertical movement in the water column, regulating to some extent their own dispersal. The length of time that larval stages remain in the water varies considerably, though two to four weeks is common amongst fish and corals. Very mobile larvae such as those of the surgeonfish can swim continuously for nearly 200 hours, covering distances of tens of kilometres. During that time these zooplankton may be moved several hundred kilometres by water currents, but there is also evidence that some species can extend this duration considerably, for example if they do not detect suitable substrate on which to settle or if food is very scarce, in which case they can enter a dormant phase until conditions improve. This phase may last for many months—some have even been referred to as being 'virtually immortal'. This has important consequences to the dispersal of reef species.

A factor important to the survival of planktonic larvae is their sensory organs. Many respond to gravity; those that are strongly geotactic will settle quickly, with the advantage that they settle near their parents and thus in a habitat that is likely to be suitable for them, but with the disadvantage that dispersal is limited. Others respond by swimming upwards, which has the converse benefit of greater distance dispersal, but it also increases the risk of moving into an unsuitable habitat—and, of course, prolongs the risk of being eaten by a plankton feeder. Many reef-dwelling species, including coral larvae, settle preferentially onto limestone and calcareous red algae in particular, using chemosensory mechanisms. Another intriguing attractant to some larvae is sound, notably the sound of the reef, which includes components of breaking waves, and the clicking and snapping noises made by reef animals; a reef-dwelling species would thereby be drawn to reef habitat as opposed to, for example, expanses of sand around the reef.

Permanent zooplankton. Except during episodes of mass spawning by corals and their relatives, the most numerous component of zooplankton on most reefs is a large group of demersal zooplankton, which occupy the water column close to the seabed. These live in crevices in the sand and rubble on the seabed during the day, and rise from the reef into the water column at night. This permanent zooplankton includes a wide range of tiny crustaceans, polychaetes, and nematode worms. Numbers of demersal zooplankton usually greatly exceed those of zooplankton that are entirely pelagic, and as a consequence demersal zooplankton are a major source of food for plankton-feeding species. Measurements of their density in sandy substrates show extraordinary numbers. Red Sea sand contains many thousands of animals in each square metre, of which three-quarters or more migrate upwards through the water column at night. Usually they do this within an hour of sunset and return to their crevices in the sand and rubble near dawn. It is this group of plankton that is the target for many filter feeding organisms: corals extend tentacles at night and featherstars and basket stars unfurl their arms to catch them: night time is their time to feed. Several major groups, again in the Red Sea, have been shown to migrate vertical distances of 25 metres from their rubble substrates to near the surface of the water, creating a steep gradient in the concentration of zooplankton over a reef. Remarkably, demersal zooplankton from many taxonomic groups show movement patterns that are extraordinarily well synchronized—emerging from the reef within a few minutes of sunset all year round, and returning to the benthos about an hour and a half before dawn. It is thought that just after sunset planktivorous fish are much less efficient at capturing zooplankton and most corals have not yet expanded their tentacles, and these factors offer a brief window of time in which the zooplankton can escape predation in that most dangerous layer just above the substrate to a metre above it, where benthic life can trap them. Presumably, predator avoidance is the main factor determining the precision of this timing.

In total, this demersal plankton is a significant component of the trophic web on a reef. Although the animals are largely unseen to the naked eye, their effects are abundantly clear in that they support much of the visible life. All these groups together form the primary conveyors of organic material through the web of life, from initial microbial decay of debris to the larger and more visible life forms.

Other microbial symbiotic connections. The key coral–zooxanthellae symbiosis has been described, but coral reefs show very many more symbioses between quite different forms of life which are also key components of material and energy transfer. Sponges provide a good example, and are viewed by some scientists as being colonies of symbionts. The sponge animal forms the framework, an ancient form of multicellular life but one that probably could not exist at all without symbionts. Most if not all shallow reef sponges contain huge numbers of symbiotic or commensal algae and symbiotic cyanobacteria in particular, harbouring these both within and between their cells. Multicellular algae, particularly the reds and greens, also form remarkable associations with sponges. But all the most important symbionts are, again, micro-organisms. Many of the bacterial and cyanobacterial symbionts fix nitrogen from the water, with important consequences to the availability of this critical nutrient. When it is considered that sponges at some depths and in some areas such as the Caribbean may occupy as much of the substrate as do the corals, it is clear that they are as important to the ecology of the reef as are corals themselves. The symbionts increase the ability of sponges to grow, to capture particles from the water column, to stabilize loose substrate, and, depending on the species, to bore into the reef and cause bioerosion. With sponges as much as with corals, symbiosis is an essential mechanism that tightly cycles nutrients and metabolic gases as well as producing higher photosynthetic products. Some sponges also benefit from certain cyanobacteria that are highly toxic. These defend sponges from predators—indeed some Caribbean sponges are toxic even to human skin.

Bacteria form common symbiotic associations in the mucous layers of corals and many more live inside coral tissues. Some of these bacteria have been cultured, but a better understanding of their diversity and abundance has been revealed with the advent of DNA examination techniques. Analysis of just a handful of coral samples has shown several hundred different bacterial species, about half of which appear to be new species, and it is thought that several thousand different kinds exist. Their diversity in any one area may parallel the diversity of corals themselves in that high latitude reefs have been shown to have fewer species than those closer to the equator. Probably as common and as diverse are members of the archaea group, microorganisms that are quite separate from the bacteria, reportedly existing in densities of tens of millions on every square centimetre of coral surface. Little is known about them or how they affect the food chains that power the reef. However, their huge numbers and ubiquitous nature suggests that they will be shown to be important.

Productivity from seaweeds. On a coral reef there are relatively few visible 'plants' such as are common on land and in colder seas, and most of the energy pathways that drive the reef system described in this chapter have involved or been based on microbial organisms. The major exception to this comes from filamentous algae, mainly green forms growing between the corals and soft corals, and these are a major food source for grazing fishes. These seaweeds grow rapidly on all areas of limestone rock in shallow sunlit waters, but there is usually not very much of them to be seen on a healthy reef because they are grazed as rapidly as they grow by the numerous species of herbivorous grazing fish, sea urchins, and other grazers that thrive on any healthy reef. Their standing crop may be small but their turnover is rapid, as can be seen if an area is experimentally caged to exclude grazers. In such sites, the substrate soon becomes covered with rich seaweed growth, mainly filamentous forms. This demonstrates significant transmission of material and energy along this food chain too. Of note is that these beds of algae, commonly called turf algae, are also a prime habitat for many other life forms already discussed, such as bacteria and protists, which are consumed by grazers when they feed on the algae.

The rate of turnover of these microscopic and simple forms of life is extremely rapid. It is not too much of an oversimplification to suggest that, although unseen to the naked eye, they power the reef. Referring to the coral veneer as well as microbial components, biologist Bruce Hatcher described it thus:

> Coral reefs are gigantic structures of limestone with a thin veneer of living organic material—but what a veneer! Everything that is useful about reefs (to humans and to the rest of nature) is produced by this organic film, which is approximately equivalent (in terms of biomass or carbon) to a large jar of peanut butter (or vegemite) spread over each square meter of reef.

Thus the quantity of living organic tissue on the reef at any one time is not necessarily very great but its rapid growth and consumption, its productivity, is immense.

Chapter 6
Reef fish and other major predators

The kaleidoscope in the water

Perhaps the most immediate visual impression for a snorkeler or diver visiting a healthy reef is the immense number of fish in a variety of shapes, sizes, and colours. Some, especially the smaller ones, are very timid, never venturing far from the protective branching corals while others, such as parrotfish, swarm over reefs in great schools scraping algae and other food material from both dead rock and living coral. On an unfished reef, thousands of fish hover in the water above the reef. No other natural habitat in the ocean shows this diversity and abundance. The diversity is so great that about a quarter of all marine species may be found on coral reefs, even though this habitat occupies only an estimated 1 or 2 per cent of the area of the earth. The exact area is hard to determine, since many reefs are deep, not mapped, and cannot be seen by satellite. Clearly, there are strong driving factors leading to the proliferation of these numerous and various species of differing size and body shape; and the variety and number of fish have great practical importance for humans, many of them providing food for the large and increasing human populations that live along tropical shorelines.

Many smaller species are coral obligates, meaning they can live only in, or feed only on, particular species of coral. This naturally constrains their overall abundance and distribution. When the corals on which they are dependent die, obligate fish species are the first to show population collapse. Although, some do survive, for example by swapping to new host corals, provided the corals contain roughly the same kind of branching structure. The association that any particular fish species has with a reef may be more or less permanent, or it may occur for only part of the lifecycle of the fish, and it might be food or habitat related.

As with the corals, there are major geographical influences on the diversity patterns of reef fishes: there are quite different assemblages in the Caribbean for a start, and there are many more species in Southeast Asia than in the Eastern Pacific. Relatively isolated areas such as the Red Sea have many endemic species not found anywhere outside that region, while some species range across entire oceans. There are patterns of diversity correlated with latitude too. Furthermore, on any one reef the diversity of the fish assemblage increases with the overall complexity or rugosity of that reef. Generally, the more complex the topography of habitats, the more complex and varied are the assemblages of fishes that utilize them. And again, as with corals, the various species have adapted to different depth ranges, with some thriving in the shallow surf zone, many requiring vast expanses of sand over which to forage, and some being restricted to deeper areas. Here, as with the corals, most species prefer intermediate depths, while the larger predators hunt across all depths.

Feeding ecology of reef fishes

The hundreds or thousands of fish species found in a reef complex show most possible feeding types, very often with one type grading into another. Our human mind prefers to categorize particular species as herbivore or carnivore, for example, but many species cross these classic boundaries. Some fish families contain species with a more or less consistent feeding habit (e.g. most shark species) but other huge families, such as the parrotfish and the wrass, contain species with everything from plankton feeders to detritivores, herbivores, mollusc feeders, and others. Examples have even been found where the same species has a different diet in different parts of the world. Fish size itself is an unreliable indicator of trophic status. Some carnivores are just a few centimetres long while the largest fish of all, the whale shark, feeds mainly on zooplankton.

Detritus feeders are very abundant. We saw earlier that the detritus element of the reef is a very important one, mediated by microbial organisms including protists and bacteria, forming the microbial loop. Detritus is concentrated on the seabed. Any hard substrate is colonized by filamentous algae, called algal turf, and this is where an enormous part of detrital processing takes place and, because it is so organically rich, it is a primary grazing site for detritus feeding fish. Detritus in algal turf may comprise up to about 80 per cent of the organic matter present, which provides more nutritional value than comes from the algae itself. From foraminifera (a phylum or class of amoeboid protist) to fish faeces, this is an organically rich site. Species of fish from several families feed on this mix and in the Great Barrier Reef these constitute about 20 per cent of the different species and nearly half of the biomass of fish in that habitat. Feeding here is based on the detrital food chain started by bacteria, and as such this is critically important. The fish that do this, such as parrotfish, are extremely conspicuous, sometimes foraging in large schools over the seabed.

Plankton feeding fish are common also. Amongst these are the smallest known: 'plankton pickers' which hover just centimetres above their bolt-hole and grabbing individual large zooplankton. Because much zooplankton commonly emerges at dusk and returns to crevices during the day, many of the fish that feed on them are nocturnal, noted for their enormous eyes. One of the smallest fish known, *Trimmaton nanus* (sexually mature at only 8 millimetres long) is an example of a plankton picker.

Herbivorous fish diversity is fairly high, making up perhaps a quarter of the total species present. Herbivorous fish can be present in large numbers and provide considerable biomass. They are mostly found in shallow, well illuminated water, since light is required by their plant food. The raspers and grazers such as parrotfish with their strong beaks take in prodigious amounts of limestone along with their organic food; this scraped limestone might be up to three-quarters of the total mass ingested and is important to their digestion. In fact, their defecation of it, after they have extracted the nutritious component, makes up much of the sand on the reef—up to one ton of this material can be produced in a year by a single, very large parrotfish, and a school of smaller parrotfish produces more than this, all from rasping at the reef. The bioerosion of the reef that this causes is therefore considerable, and these rasping fish have a major structural function on a coral reef. The scraped areas are quickly colonized by benthic life, including, crucially, larvae, which will go on to form a new generation of corals.

Since the algal turf contains so much foraminifera and protists, the place in the food chain of the fishes feeding on them is certainly not one of clear-cut herbivory. Some fishes favour certain groups of algae while others are more generalist. Some, such as several damselfish in the large Pomacentridae family, become very territorial over favoured patches, vigorously repelling other encroaching fish as they try to keep a few square metres for themselves. Their behaviour is very much linked with their diet. Once, when I staked out quadrats for coral measurements using steel pegs, damselfish promptly made several stakes the centre of their own guarded patches so that, on returning three months later to measure coral cover and recruitment, I found damselfish were 'farming' these experimental areas, which were now smothered with algae. Indeed they even tried to repel me—these little fish are instinctively fearless, even though they are easily overwhelmed by, for example, a school of parrotfish as it charges through the area.

Human-produced sewage or agricultural enrichment also triggers a lot of algae growth on a reef, however, many of the resulting algae are unpalatable and not eaten by fish, so the algae continue to grow to the extent that they exclude the corals they grow over.

Some herbivorous fish, like the detritivores, forage over large areas, some showing migration patterns between reef and adjacent sandy areas, perhaps on a diurnal basis or perhaps with cycles linked to reproduction. Although turnover of algae, namely its growth and its consumption, may be very high, at any one time the standing stock, or biomass, is generally low because of continued cropping.

Fish that feed on invertebrates include generalists that will eat most things they can find. Other fish, however, are more specialist. Feeding on invertebrates that live just under the surface of the sand requires very different skills and feeding equipment to fish that chomp on dead coral rock to obtain the invertebrates living within, which leads to very different adaptations. Goatfish, comprising the Mullidae family, feed on sand using a pair of sensitive barbels ('feelers') beneath the chin with which they can sense their prey. Many have a specially shaped mouth with which to get at their prey in rubble. Rays have mouths under their heads and use their wings to disturb the sand to shake up their prey. Some fish feed on sponges, which also requires specialized digestive processes to cope with both the sponge's toxic chemicals and the sharp siliceous or calcareous texture of the sponge matrix.

Some fish feed on corals (i.e. corallivorous). The very attractive butterflyfish is a classic example, and these make up about half of the estimated 130 species of corallivorous fish known to exist. Some butterflyfish exist exclusively on just one or two species of corals and have developed elongated mouth shapes to facilitate picking off individual polyps. About one-third of corallivorous fish feed only on corals, obtaining nutrients from polyps and coral mucus, while the remainder also feed on other invertebrates when these are encountered.

Carnivorous fish at the top of the food web may be the most numerous in some areas of reef. Many detritivores are at least partly carnivorous too, ingesting small worms, crustaceans, and other small animals as they forage. Some unexpectedly docile but large sharks, such as the nurse shark (Figure 12) feed on crustaceans and molluscs. But the so-called piscivores, the fish hunters, sit at the very top. These are generally the fastest swimmers as they have to catch such fast moving prey. It is thought that the common tendency of schooling by many potential prey species of fish is because schooling helps to either confuse the hunter or because it provides a form of safety to individuals—much as with herds of wildebeest in Africa. However, some predators school too: some barracuda and jacks hunt in coordinated packs, and a classic sight on a reef is schools of very fast swimming jacks, barracuda, and others, which 'carve

12. A large nurse shark, a predator of molluscs and crustations

up' the schools of prey species, isolating some individuals before snapping them up.

The modes of operation of piscivorous fish, and their range of body shapes, are as varied as their prey, and include stalkers and ambushers, as well as the active, sleek, and fast hunters. Some predators like the grouper simply lie in wait and gulp passing prey by opening their mouth rapidly, causing an inrush of water and, with it, the hapless target. The piscivorous sharks are amongst the most visually impressive hunters on the reef. They have astonishingly acute olfactory senses and organelles that detect vibrations and electrical signals in water, including those coming from an injured fish. Sharks have an important 'top-down' controlling influence on reef ecology, as do many other large piscivorous groups, but their place in our folklore is unparalleled because of the very occasional attack by some species on people; usually these attacks have been on swimmers rather than on divers, except in cases where the latter have been spearfishing and are hanging on to injured and dying fish.

However, sharks are currently experiencing appalling destruction at the hands of humans in all oceans of the world, sometimes for consumption of their flesh but often, more wastefully, just for their fins. Consumption of shark-fin soup is enormously popular in Asia. One hundred million of these high-level predators are taken every year, with enormous ecological consequences. Unfortunately, the unfounded perception of the danger sharks pose to humans is not really changing. Ignorance about sharks and their important role is remarkable: even at a marine science conference in China, I was served shark-fin soup (and my refusal, as a guest of honour, caused bafflement as much as consternation). About a hundred shark attacks on people are recorded each year (one millionth of the number of sharks being killed) and about a fifth of these are fatal. Most of these attacks occur in cool water areas and estuaries—not in reef waters.

Piscivores can feed on other carnivorous fish as well as on herbivorous fish, so they cannot be conveniently placed into one simple level of the food web. In fact there is no single classical structure of proportions of fish on a reef. Taking those reefs mentioned earlier that have highest fish biomass, in Kingman reef in the Pacific, the apex predators are mainly sharks; in Chagos in the Indian Ocean, on the other hand, the large apex predators include far greater numbers of other groups such as snappers, jacks, and groupers, with far fewer sharks. It is not known to what degree this is the result of illegal fishing and poaching, which undoubtedly occurs in both places, all such fish being highly desired by humans for food. And where there are human settlements nearby, there is invariably the greatest reduction in these top-level carnivores, which swiftly causes ecosystem distortion. The removal of these carnivores might be expected to lead to 'predator-release', in other words, an increase in numbers of their prey, such as herbivores. However, the herbivores are also fished in large numbers.

Overall, while a healthy coral reef does have, like any ecosystem, a roughly pyramid shaped trophic structure, the fish component alone does not necessarily have this structure. Indeed, the 'fish pyramid' on its own can even be upside down. However, one important point about a coral reef is that, as with plant biomass, it is not only the biomass per se that is important but the productivity, or rate of turnover. As with all groups, rapid growth and rapid consumption may lead to low biomass—and a reef does demonstrate a very fast pace of life. A second important point is that because fish are mobile many, including many of the numerically abundant species, roam over much larger areas than just the reef itself; they may travel over large areas of adjacent seagrasses and sand, or they may pursue shoals of prey into deep ocean water. This means that fish form major connections between reefs and adjacent habitats.

The complicated ecological structure of fishes on a reef is clarified by the use of nitrogen isotopes to determine the trophic level of any particular fish. The trophic level is a measure of how or how much it feeds, with plants being level 1, herbivores level 2, carnivores level 3, and higher carnivores that eat those carnivores being level 4, etc. You can most easily see what a fish has recently been eating by examining its stomach contents or by simply watching it. However, such observations are usually too brief to show its typical diet. So the ratio of different stable isotopes of nitrogen is used. This changes very slightly with each step along the food chain. Most nitrogen is in the form of ^{15}N (nitrogen with 15 protons and neutrons in the nucleus), but ^{14}N (with one less neutron) forms about 0.4 per cent of nitrogen in air; this ratio changes very slightly with each step in the food chain. The $^{15}N/^{14}N$ ratio increases up the chain such that herbivores have a different proportion of ^{15}N that do carnivores, for example. Sensitive measurements of the ratio of nitrogen isotopes have been used to determine that many, if not most, fishes' trophic level is not an integer, meaning it is not entirely a herbivore, or a primary or a secondary carnivore, but has a trophic level of, say, 1.5 indicating that it may be mostly a herbivore but consumes small invertebrate animals too, or it may be 3.6, meaning it is a predator of primary carnivores but it also takes in a good proportion of higher level carnivores as well. The method has also shown that some herbivores remain at the same trophic level throughout their life (i.e. eating the same plants), but that some carnivores change their diet as they grow larger, feeding on progressively larger carnivores at higher trophic levels. Such methods of observation and measurement give a much more accurate picture of how a coral reef works.

Marked interactions have been seen between fish and the corals themselves in all oceans of the world, but other animals have major controlling functions too. Some examples are easy to understand: if we eliminate fish which feed on grazing sea urchins, then the sea urchins will increase substantially; urchins graze by rasping on the coral rock, and the cascading effects might include much greater reef degradation. Studies in East Africa and the Caribbean have shown how removal of triggerfish released pressure on their sea urchin prey and led to an increase in urchin numbers. This led to reductions in algae as well as reef erosion. In the Caribbean, there was a well researched, massive mortality of urchins from a disease thought to have reached the reef via the Panama Canal, which all but eliminated the dominant grazing sea urchin *Diadema*. The cascade of effects that followed included an increase in seaweed at the expense of corals. The concept of predator release of prey is particularly significant on coral reefs in the present days of overfishing, and the explosion in algae growth that may result could swamp a coral reef. There is nowhere quite like a coral reef for complexity of possible effects to any given change. One example of this is seen with the explosions in numbers of another major coral predator, the crown of thorns starfish.

Crown of thorns starfish. The colourful crown of thorns starfish, *Acanthaster plancii*, is a notorious coral reef predator. It is large, up to about 40 centimetres in diameter, with eight to 21 arms, covered with poisonous spines, and comes in a variety of striking shades of blue and red. It is a very efficient predator of corals. Plagues of these starfish have appeared on occasion, always unexpectedly, and in the course of about a month they can consume more or less all the coral over hundreds of hectares of reef. They feed by hunching themselves over a large, relatively flat boulder coral or by curling themselves around the branch of a staghorn coral, extruding their stomach to dissolve all coral tissue beneath and sucking back the digested contents. Then they move to the next patch. In such plagues, there may be several starfish in each square metre, and total plague populations can number tens of millions.

Their natural range extends from the Red Sea in the west through to Panama in the Americas, but they do not occur in the Caribbean region. An occasional adult crown of thorns starfish is a natural sight on most coral reefs. However, the earliest known plague outbreaks were first recorded in Japan in 1957 and on the Great Barrier Reef in 1962; outbreaks have been reported increasingly frequently since. For some reason, every now and again they appear in vast numbers, moving on to a reef from deeper water, consuming just about every coral they can get to (and showing some cannibalistic behaviour for good measure). Then, when there is absolutely nothing left to eat they disappear equally suddenly, leaving behind a spectacular white expanse of bare coral skeletons. A week or two later these skeletons become darkened as they are colonized by filamentous algae, and the dead corals become subject to countless burrowing, tunnelling, and boring animals which quickly exploit this new habitat.

Recovery by corals from an outbreak can occur, although it does take years, as corals grow relatively slowly. Coupled with other pressures on the reef, these crown of thorns starfish greatly reduce reef stability and persistence, and at times even its very existence as an ecological entity for years to come.

There may be several causes of these crown of thorns outbreaks. Increased nutrient flow, which provides planktonic food for the animals' early life stages appears to be one factor, at least in some places, and it is relatively easy to measure increased sewage discharge or runoff of nutrients from agricultural land. Juveniles remain cryptic and feed on coralline algae initially but switch to corals when young adults. Other postulated causes of population explosions have been overexploitation of some of its predators, including the large Triton gastropod molluscs, much sought after for their attractive shells. An intriguing thought is that there is a threshold in density that needs to be reached, a density perhaps above which exchange of hormones or signalling chemicals somehow triggers an explosion in numbers and a migration from deeper water onto a reef. Many control methods have been tried after various outbreaks, but in most cases these measures have been too little and too late. Simply cutting up a starfish makes the situation worse because multiple adults develop from the original.

Preferred methods now include simply collecting them by hand or injecting them with toxic chemicals, but these solutions are very labour intensive and inefficient. The fact that they are used at all reflects the urgency, even desperation, that a plague can engender for humans. As a result, these methods might be attempted near an economically valuable tourist site, for example, but they would have no impact on, for example, a wave of starfish travelling along an area the size of the Great Barrier Reef. Predicting plague outbreaks remains impossible too; their larvae for example can travel thousands of kilometres and, to add further complexity, it is now thought that there might be several very similar species, rather than just one.

Fish commensals

The variety of ways different species of fish have evolved to live together or with completely different invertebrates to mutual benefit is countless—and intriguing. Some seem weird to us: certain species of the attractively named Pearlfish have a somewhat less attractive habitat of living inside sea cucumbers, entering and exiting through the sea cucumber's anus. Shelter is the main benefit gained, and it seems unlikely that the sea cucumber obtains reciprocal advantage. The relationship between gobies, a type of small fish, and shrimps, which share a burrow, seems rather more appealing to us. The shrimp digs the burrow and, while it looks as though the shrimp does all the excavation work, the fish stands guard, flicking its tail to warn the shrimp of danger.

But perhaps the most iconic symbiosis on the reef is that between clownfish and sea anemones. In the Indo-Pacific, but not in the Caribbean region, 28 anemone fish species form symbioses with about ten different species of anemone (Figure 13). Some of these fish are specific to one species of anemone while others will adopt any or all species of anemone as a home. The anemones have stinging cells and it has long been a mystery how these symbiotic

13. Anemone fish, often called clown fish because of their striking colouration, living in a sea anemone in a remarkable symbiotic arrangement

fish avoid being stung and killed. Several mechanisms are postulated: the fish might exude a chemical which suppresses the triggering of the stinging cells, or the constantly flickering, agitated activity of the fish might transfer chemicals from the anemone onto the fish's body, so that the fish is sensed by predators as being part of the anemone—after all, one tentacle does not sting another. In return for protection, the anemone benefits from more food, including that from the fishes' excretory material. The latter is certainly likely to benefit the anemone's zooxanthellae; all anemones showing this association have zooxanthellae too. The presence of anemone fish is also known to lead to an increased proportion of time in which the anemone's tentacles are expanded and thus functioning, so there are multiple benefits to both.

A final charismatic example comes from symbiosis between different species of fish. At 'cleaner stations' on a reef, larger predatory fish line up to have the parasites infesting their skin, gills, and even their mouths removed by various species of cleaner fish. Such cleaning stations are commonly at fixed locations on a reef, where a fish signals its desire to be cleaned by adopting a special posture, an invitation to receive attention from the cleaner. The little cleaner fish restrict their diet to dead tissue or parasites, and the predators do not eat the cleaners. However, rarely is a trick missed on the reef: another fish which looks like the cleaner, with the unlikely name of sabre-toothed blenny, *Aspidontus taeniatus*, is a cleaner mimic which will similarly approach the large fish waiting at the station, but instead of helpfully cleaning it, the blenny will take a chunk out of the larger fish's tissues, such as its gills. Naturally, this deters the large predator, and provides a fascinating example of population balancing because the number of sabre-tooth blennies can never be too high or else no large fish would come to be cleaned again.

There are many other examples of symbiosis, reflecting the high diversity found on a reef.

Reef fish biomass, productivity, and fishing

Given the importance of reef fishes to so many human communities, an important question is: what is the productivity of the reef habitat in terms of fish, or, how many fish can be extracted without compromising the system? Because reefs have been already so heavily fished, it is difficult to obtain accurate estimates of what an untouched reef might actually look like. One measure for this is reef fish biomass.

In one of the least affluent regions of the earth, the Western Indian Ocean, there have been some areas that have been protected from fishing for several decades and it has been suggested that the reef fish biomass there might reflect the true level of life a well managed reef might be able to support. But a recent study of historical fish catches between the 8th and 14th centuries, carried out by examining Swahili kitchen middens, has shown that the species' composition several centuries ago cannot have been the same as it is today. Even in relatively well protected reef areas in Kenya, for example, the size of adjacent human populations is such that reef fish composition has drastically changed over time. Although such sites are protected now, some of the best managed reef areas still show marked changes and depletions of fish populations, so that they cannot really tell us what the fish population in a 'pristine' reef might be. How much fish might have existed under the best hypothetical conditions millennia ago?

Several researchers have assessed reef fish biomass in a standardized way across the world, focusing on the potentially very rich zone at 9 metres deep. Most places measured have a biomass of reef fish of about 1,000 kilograms per hectare or less, and very often, as in the case of particularly mismanaged reefs, only about 200–400 kilograms per hectare. The best managed marine protected areas in the Indian Ocean with people nearby have a biomass rarely more than about 1,500 kilograms per hectare. In many countries, areas of reef that are designated as being protected in fact showed no greater fish biomass than designated fishing areas, which clearly suggests that such areas are protected in name only. The same situation holds across the Pacific as well. Only very few locations in the world have been found where the reef fish biomass dwarfs that of even the best managed protected areas near inhabited places. Richest of all is the large Chagos Archipelago which has not been exploited for reef fisheries for many years, and where the biomass at 9 metres depth is over 7,500 kilograms per hectare, six times that of the largest biomass found elsewhere, and a hundred times or more greater than some of the very overexploited reef systems in this ocean. The reason appears to be that it has been mostly uninhabited for the past 40 years, it is very remote, and quite simply it has suffered from very little fishing. It is large, being a network of several atolls and offshore banks, the whole forming what is the world's largest single expanse of virtually untouched reef. Similarly high biomass has been found in the other two major ocean basins also: in Kingman atoll in the Pacific and Cozumel on the Caribbean side of Mexico. Tellingly, in both these places fishing is very low too. In the Hawaiian island chain, biomass values were around 500–1,000kg per hectare in inhabited islands, rising to 3,000, and in one particularly remote area there was a biomass of 4,500 kilograms per hectare, the higher biomass locations having the most apex predators (like sharks or groupers) of all the locations measured.

There are differences between the three sites with the highest biomass in terms of the kinds of fish that make up the biomass. Kingman and the richest Hawaiian site, for example, have a very high biomass of sharks and top predators in particular, much more so than Chagos, which has a higher biomass overall. In any case, the message is depressingly clear: where there is any fishing at all the fish biomass goes down severely; it takes very little fishing effort to cream off the riches.

One aspect of fish physiology is crucially important to replenishment of many fish species: the larger an adult female fish is, the more eggs it will produce. This is possibly an obvious point, but, importantly, the increase in egg production is not linear. To use a hypothetical example, a single 10kg female might produce many millions of eggs per year, while even ten 1kg females combined of the same species would produce only a few thousand per year. If we remember that the larger fish are the most prized in the fishing industry, we can immediately see that the damage done to the ecosystem by removing the largest fish is exponentially greater. I would stress here that no blame should be attached to those fishing at subsistence level for collecting what they can; for these people, it is usually a matter of survival. Leaving this consideration to one side, it is nevertheless a salutary point to note that even a very modest level of fishing intensity can cause much ecosystem distortion very quickly. In areas that were once protected but which then permitted fishing, ecosystem collapse happened in only a very few weeks.

Fishing is an emotive area because of consideration of the needs of those who live at subsistence level, but we must understand the situation thoroughly if we are to try to do anything about the overfishing problem. It is demonstrably no good claiming that local people fish sustainably, as is so often claimed. I have challenged many people to produce examples of where an active coral reef fishery is even remotely sustainable, supporting a high biomass, which is maintained at that high level. On the other hand, there are countless examples of 'sustainable' production that is at a greatly depleted level. The importance of restoring fish abundance and productivity is growing now that the issue of food security has become so prominent.

Some of the above studies were carried out partly because fishing is thought to be one of the most pervasive threats to coral reef maintenance, due to the important function fish have in regard to the coral reef itself. Because fishing the reefs is so essential to human survival and nutrition it has been accepted as being an essential pursuit, a necessary evil, as it were. There is, as a consequence, an exploitation gap: the gap between the high biomass/productivity areas we see, for example in Chagos, Kingman, and part of Cozumel, and the low biomass areas (which lack large, reproducing adult fish) nearly everywhere else. It is a curious fact that, as far as I know, almost no area has been shown to have fish biomass levels of between 2,500 and about 6,000 kilograms per hectare. With, to my knowledge, one recorded exception in the Hawaiian chain, there are only those with the higher biomass figures, which are now very rare, and the almost universally low biomass figures seen where reefs are exploited. We will explore the likely reasons for this gap later in the book.

Chapter 7
Pressures on reef ecosystems

A fundamental character of a coral reef is the enormous number of interlocking connections between components, a consequence of extremely high biodiversity. As the number of species in any ecosystem increases, so does the number of connections between them, and connections rise at a rate much faster than the number of species. Connections can be destroyed, as can the species themselves. The question of whether a reef ecosystem is more robust to outside pressures as a result of this huge number of species and connections, or made fragile because of them, has been endlessly discussed. The answer must be 'both', depending on what reef system is being discussed, its geographical location, and what the pressures are.

Coral reefs today: uses and abuses

The uses to mankind of coral reefs, sometimes called 'services', have been numerous ever since people first settled beside them thousands of years ago. They supply food from fishing, coastal protection, and shoreline stabilization. Many countries are composed of reefs or are partly protected by them, and some of these, such as some in Southeast Asia and several Indian Ocean and Caribbean islands, are extremely populous in the parts that support coral reefs.

Reefs composed of similar corals as those we have today have existed since the Triassic Epoch, sometimes developing massively, at other times existing in a less prolific manner up to the present time. In the Quaternary period, glaciations have come and gone, and reefs have survived them. The time since the end of the last glaciation is known as the Holocene. But recently it has been proposed, by serious scientists, that the time from which human impact on the environment has become significant, even in geological terms, should be renamed the Anthropocene. Humans have, for example, now moved more surface sediments than all of the Earth's rivers and glaciers during the last ice age. There is some disagreement, probably trivial, about when the Anthropocene epoch should start. Some prefer to date it from the beginning of the Industrial Revolution, others suggest as recently as only a few decades ago when our climate started changing noticeably. Some have said it should start as far back as two millennia ago, because already substantial changes were taking place to the Earth's vegetation as a result of human activity, when great ancient cities developed, flourished, and then were abandoned when their populations outgrew the ability of accessible habitats to support them. If this seems exaggerated we need look no further than the huge accumulations of turtle and mollusc shells piled deep in many coastal locations such as Turtle Mound in Florida (actually about 30,000 cubic metres of oyster shells) and turtle shell middens in several Arabian countries such as Oman and some islands in the Red Sea, testament to the enormous quantities of seafood that was being extracted thousands of years ago.

Whatever the date of the start of the Anthropocene, the point is that changes being wrought by humans in much of the oceans, especially in the tropics, now clearly match and in many cases exceed anything that reefs have experienced in past human history, and they match or even exceed the changes caused during past extinction events in geological history. Many other major habitats on the Earth are probably in a similar position, but coral reefs have for a long time been considered to be the litmus, the canary in the coal mine, which foretells substantial and deleterious changes happening to the Earth system. They are a major ecosystem that can be viewed, in the modern scientific sense, to be suffering unwise use to an extent that now threatens their survival. This is not an exaggeration: about a quarter of reefs that existed 50–100 years ago are now largely dead, and another quarter or even third (depending on the area) are in critical condition; only about a quarter of the remaining reefs are in a healthy, robust state.

There are various causes for their decline, and they can be grouped into two main categories: those caused by numerous, usually localized, impacts, such as different forms of pollution, overfishing, and shoreline developments that create large quantities of sediment; and those caused by the much more recently recognized and probably more serious, long-term factors associated with climate change. All of these are ultimately interlinked and all are caused by human activities. A further problem comes from the fact that many of these factors occur concurrently, each one exacerbating the harmful effects of the others. It would be bad enough if these effects were simply additive, but their effects in combination are commonly multiplied: their effects may be synergistic.

Local pressures

Sewage and runoff. A major local effect, probably from time immemorial, has been that of nutrient enrichment. Coral reefs are well adapted to living in nutrient poor conditions, but sewage enriches the marine environment with nutrients, especially compounds of nitrogen and phosphorus. Sewage is a fertilizer for plants, so such discharges commonly lead to vigorously increased growth of fleshy algae. Unfortunately, these algae commonly include species unpalatable to grazers, so algae thrive and out-compete the much slower growing corals and soft corals. Seaweeds take up space, they grow faster than corals, and they shade them. Over the past 50 years, countless reefs have been destroyed this way, so that many former reefs are now merely limestone platforms covered with algae and devoid of corals. Furthermore, such pollution can also inhibit corals directly, because phosphate has an inhibiting effect on the coral's calcification process. Phosphate concentrations above just 1 micromolar solution will reduce its calcification and growth.

The stimulation of plankton numbers by sewage is equally important: more plankton in the water means that less light can penetrate to the seabed for photosynthesis. Other linkages are being discovered as well: the crown of thorns starfish discussed earlier is a case in point, in that increased nutrients, whether from sewage or agricultural runoff when too much fertilizer is applied to fields, stimulates phytoplankton, which may stimulate this predator's larvae.

*Landfill, dredging, and sediment*s. Landfill and dredging have localized but extremely severe effects on coral reefs. At its most obvious level, landfill simply buries the reef, and there are many examples where the coastal strip offshore, formerly supporting perhaps a continuous fringing reef or perhaps a series of patch reefs, is now buried under concrete because of construction schemes. Fringing reefs are flat, very shallow, and by the seaside, so they provide highly desirable real estate when built upon. It is an unfortunate fact that the price (not value!) of a hectare of coastal reef increases enormously when it is no longer reef (the same also applies to seagrass beds and mangroves). This is not whimsical: about half of the Arabian side of the Arabian Gulf is now artificial, with enormous consequent loss of all marine habitats including coral reefs, seagrass beds, and mangroves, all of which are nursery grounds for marine species. Many other reef locations are being similarly developed.

One major problem, which exacerbates this, is that of perception and economics. A simple matter like terminology is very important and allows coastal engineers, for example, to refer to 'reclamation' when in fact they are not 'reclaiming' anything at all but trying to 'claim' an existing marine ecosystem. The terms 'landfill', or 'reef-fill', would be more accurate in these cases. Underlying this problem is simplistic economics. The tyranny of the spreadsheet means that a development company, for example, or a harbour authority, costs only its own immediate activities—their own labour, concrete, and equipment used to convert living habitat to concrete building foundation (they would use the term 'develop' rather than 'convert', of course). What is omitted from the calculation are the wider costs, such as maintaining nursery grounds and an adequate food supply for marginalized villagers; or the cost of removing shoreline protection, which is something particularly important to many tropical countries. Vast areas of land are now being created by these means in many tropical countries. One colleague from Bahrain describes the country as being the only island that no longer has a coastline, meaning that almost none of it is left in its natural state. The trouble is, humans do need that natural state for many different 'services'.

Such landfill commonly has a triply bad effect because material used to level and fill it may be taken from adjacent habitats (commonly misnamed a 'borrow pit', without the usual sense that 'borrowing' has of 'returning it later'), thus the adjacent habitat is destroyed as well. Furthermore, while this is going on sediment plumes invariably can be seen extending for kilometres down-current of the dredging, smothering yet more distant habitats with silt. Corals are particularly sensitive to sedimentation. Pollution control methods for such activities do exist in engineering terms, such as deployment of screens around cutting heads and dredging units, but they are expensive and usually the first kind of activity to be sacrificed to cost saving. The so-called footprint of such tropical shoreline-altering activities invariably extends far beyond the development itself, even though, today, most such developments are supposed to be preceded by some kind of environmental impact assessment followed by mitigation procedures.

It is not easy to quantify the effects of sediments on corals although this is frequently attempted. Responsible engineers are keen to find a simple measurement of what will cause harm to a reef, but this is difficult because there may be several hundred species of corals and soft corals making up the reef, each with widely varying susceptibility to sediment and with different responses to the light reduction caused by sediment in the water. The two impacts act synergistically. Sediment can be actively shed to a certain extent by most corals but this requires energy expenditure, which comes essentially from its photosynthesis just when the shading of light by sediment in the water is greatest. Colony shape is important too; open lattice branched forms do not accumulate sediment so suffer less from sediment than, for example, leafy forms. We know that the water bathing a reef is usually clear, with corals typically adapted to about 5 milligrams of sediment per litre or less. However, shoreline excavation may increase this ten-fold; and a mere four-fold increase can cause mortality to many species of corals. While all species can tolerate brief exposures to high levels, such as might be caused for example by natural events like storms, prolonged exposure to greatly raised levels for the duration of a major construction event has often killed several square kilometres of coral reef.

There are many more impacts of this type that may be locally significant. Anchors from boats cause heavy destruction, and in some of the popular recreational cruising grounds such as in the Caribbean, anchoring has destroyed entire reefs along moderately sized islands. Quite simply, the anchors drag, and even when they don't, the boat swings about its chain, grinding down the corals into rubble. Even small recreational yachts can kill a circle of coral reef several tens of square meters in size, with each anchor drop.

The provision of mooring buoys would provide a simple solution to this problem, but their use is far from universal.

Chemicals, metals, and oil. The range of industrially derived chemicals discharged into the sea is growing all the time. Many of these chemicals, toxic organic compounds and metals such as lead, mercury, and several other metals used in industrial processes, are directly toxic. However, even when their levels are very low, they can accumulate higher up the food chain so that top predators may be killed or suffer physiological impairments. Effects range from loss of fertility, slowed growth, and, in some cases, more bizarre effects like sex change. Some are accumulated in fatty tissues and others might affect neuronal systems or liver function, for example. The range of effects is enormous and this subject has been extensively researched for many years. It does seem, sadly, that very often this research does not lead to clear controls or changes in legislation in many places, and too often work on this has been little more than a postmortem, carried out to see why a reef system (or other marine habitat) was killed.

Pesticides may have disproportionately large effects. After all, these chemicals are designed to kill, so severe effects should not be a surprise. In reef corals pesticides may inhibit zooxanthellae for example, so that even if the animal part of the coral–algal symbiosis is not directly affected, they are affected indirectly through the loss of the symbiotic partner. The main source of these pesticides has long been the runoff from nearby agricultural land that has been treated with these substances. A new and increasing source of pesticide is from fish farms and shrimp ponds, where biocides of various kinds are applied vigorously to prevent disease transmission amongst the unnaturally high densities of animals that are kept in such enclosures; many of these biocides then leak on to nearby reefs.

There are many exotic and sometimes unexpected effects from chemicals. Recreational marinas contain pleasure craft and are a source of significant quantities of toxins from antifouling paints used on yachts. These chemicals are designed to kill marine life with the intention of preventing fouling on boat hulls so, not unexpectedly, they do exactly that. One recent example is a 'booster biocide' which inhibits the photosynthesis of the zooxanthellae in corals at exceptionally low concentrations of only 50 nanograms per litre, or 50 parts per billion. Naturally, it makes very effective antifouling paint.

The problem of pollutants is compounded with that of sediments noted earlier. Sediment particles adsorb toxic compounds onto their surfaces and concentrate them. Filter feeding animals which feed on these particles therefore ingest not only the sediment particle instead of a food particle, but also take in a packet of concentrated, adsorbed toxic substances as well.

Oil pollution usually hits headlines when it happens, but oil has complicated effects on coral reefs. It is commonly held that oil and water don't mix, but that is only partly true. There are many substances in oil which are water soluble, and some of these are highly toxic to most marine life. In the Arabian Gulf the oil that is produced is a fairly light crude, and the many oil spills that occur in the Middle East cause limited damage; the oil simply floats over the top of submerged reefs without doing much harm. (This is not to be confused with the effects of *any* kind of oil on shorelines, where it accumulates, leaving thick tar.) This can be contrasted with some Caribbean oils, which have notoriously damaging effects on coral reefs. In one famous oil spill in Panama near the Smithsonian's Galeta Island research station, mortality was very heavy and extended many metres below the ocean surface. The difference in the proportion and identity of some of the organic substances in oil, especially water-soluble components, lies at the heart of different oil toxicities.

Frequently, pollutants are discharged in pulses, so that, although the average quantity over a year might have been tolerated by coral reef organisms, massive, episodic pulses commonly tip them over the edge. Water conditions after a pulse may quickly return to normal, but the damage will have already been done.

Then there were nuclear weapons tests. From the mid-1940s to the early 1960s many weapons were tested on certain atolls in the Pacific. Atolls were an extraordinarily bad choice for such testing since they are amongst the most diverse ecosystems on the planet. The surface and sub-surface explosions heated surface water to 55,000°C and created shockwaves 30 metres high, with shock columns extending to 70 metres deep in the lagoon, travelling at 80 metres per second. Entire small coral islands vanished, and coral fragments were flung onto the ships observing the explosions. In Bikini Atoll especially, where considerable work had previously been done on the taxonomy of its corals, a survey done 50 years later, in 2005, showed that more than a fifth of the original coral species had not yet recolonized from other areas.

Spectacular for its scale and destructive effects on atoll sized reef systems, weapons testing on atolls eventually did stop, although whether their cessation had anything to do with the biological richness of the sites is debatable.

Coral diseases

In the last couple of decades many coral reef diseases have been identified. It is difficult to culture most of the disease organisms—which is the usual first step in identifying any disease—so it has rarely been possible to determine exactly what many of the pathogens are. As a result, some researchers have used the word 'disease' only when the pathogen is known, but 'syndrome' when it is not. A syndrome may become a disease when research determines the identity of the pathogen.

Probably the first documented coral reef disease was of Caribbean sponges in 1938 when 70 per cent of individuals in examined areas died from a fungal infection. Since then, reports of disease have increased. The increase in incidence of disease has been linked to deterioration of water quality. Many diseases and syndromes are associated with multiple infecting organisms: bacteria, protists, and minute animals. Viruses have also been implicated. With many diseases, however, it is unclear which of these might have been the original pathogen and which simply appear immediately afterwards to feed off the newly available necrotic coral tissue. There is even a tiny crab about a millimetre or two across associated with disease patches on table corals, and even where dozens are seen on a coral it is unclear whether the crabs, which are clearly eating the coral tissue, cause the disease or are simply following the band of dead tissue killed by another infectious agent.

Sewage discharges and runoff from farmland containing animal faeces contain numerous pathogens, some of which are clearly associated with coral reef diseases. Most names of diseases simply reflect their appearance: white band disease, black band, white pox, yellow blotch, and so on. This seems little more advanced than medicine in the 18th century! White band disease in the Caribbean has had critical and spectacular effects, especially on branching *Acropora*. The area of killed tissue spreads quickly along the limbs of the giant elkhorn coral, *Acropora palmata*. The white band refers to the denuded and exposed white coral skeleton, and it appears to move because as the disease progresses along the limbs of the coral, it exposes the skeleton, which stays white for a week or two before being overgrown with filamentous algae, darkening it again. Meanwhile the band of exposed white skeleton moves on. Elkhorn coral used to form by far the largest branching coral structures in the sea, and in the Caribbean it developed vast, impenetrable thickets in the shallowest areas, creating entire forests not replicated anywhere else. But white band disease removed most of these coral colonies in the 1970s.

Some elkhorn coral remains, but today nothing like the original populations can be found anywhere, and the disease itself still exists. The causative agent might be a species of *Vibrio* bacteria, but this is not certain—the epidemic came, killed, and disappeared too quickly for researchers to really find out. A later disease affecting Caribbean corals has been termed white pox disease. It is equally contagious and is manifested by white patches that spread at speeds of up to 2 centimetres per day^{-1}, with patches merging to kill the entire colony. White pox is caused by a bacterium, *Serratia*, a coliform found in sewage from both animals and people. Whereas white band disease all but eliminated this most important of corals, white pox keeps populations suppressed. Both diseases also affected thickets of the branching staghorn coral *Acropora cervicornis*.

Caribbean sea fans, or gorgonians, are equally conspicuous on reefs, especially in the Caribbean. Many are the size of a diver, or bigger. Caribbean sea fans have been affected by the fungus *Aspergillus*, which, in the last 20 years, has caused the death of increasing numbers. The disease appears as patches of red-brown colouration that expand, darken, and then kill the entire colony. This fungus occurs naturally in air and soil, and its concentrations in rivers have increased enormously with deforestation and agriculture so that now the pathogen is carried throughout the Caribbean from the Orinoco and Amazon rivers, amongst others. A striking link has also been demonstrated with aerial introduction of *Aspergillus* spores from soil swept up over North Africa and deposited in the Caribbean region. The dust has in addition been linked with increasing respiratory diseases in humans. Hundreds of millions of tons of the soil, carrying pathogens, is deposited into the Caribbean during periods that seem to correlate with outbreaks of this disease amongst sea fans.

The sea urchin *Diadema* is widespread, but those in the Caribbean suffered a spectacular die-back in the 1980s. This urchin is one of the most important herbivorous grazers on Caribbean reefs. The first widespread mortality was noticed off Panama, and a wave of mortality then spread out from that point following water currents, so that almost all Caribbean adult *Diadema* were killed within a couple of years. Curiously, juveniles mostly survived, and populations are recovering today. The loss of these sea urchins was likely to have been a cause of, or at least a major contributor to, the increase in algae now found throughout the Caribbean. Increasing nutrients from sewage and farmland, coupled with overfishing of herbivores were important too; it may be futile to try and determine which factor is the most important because it is clear that all of them have acted in concert.

Caribbean reefs appear to suffer more from disease than is evident in the vastly larger Indo-Pacific, which may be due partly to the greater amount of research done but partly also because the Caribbean basin is very heavily used by large numbers of humans. But diseases are also known and are increasingly researched in the Indian and Pacific oceans too. There have not been such widespread outbreaks of disease in these oceans, perhaps because of the more dispersed nature of the reefs, though this is speculative and the Indo-Pacific has experienced vast and equally destructive plagues of other predatory animals, notably the crown of thorns starfish.

I have explained at some length the pollution and disease problems of coral reefs for two reasons. First, reefs support a greater diversity than anywhere else in the oceans. They also support, through their potentially very high productivity, greater numbers of people per unit area than any other natural ecosystem. The severity of impact by humans on reefs has a disproportionately high effect on us. Obviously it is true that, for example, deforestation of several hundred square kilometres of Amazon rainforest greatly affects the local people there, but the extent and consequences of destruction of coral reefs can be argued, from several standpoints, to be equally or even more severe. Furthermore, reefs are the oceans' early indicator of their environmental state.

One other factor may have severe consequences—they certainly have striking effects—and that is invasive species. An invasive species is one that is not indigenous to the area, perhaps introduced in ballast water from tankers or by an accidental release from aquaria. Most introduced species don't do much damage, but some find a vacant niche and their numbers increase hugely, at which point they are termed invasive. One of the most striking examples of all is the Indo-Pacific lionfish, which was accidentally introduced into the Caribbean in the early 1990s. Thought to have originated from aquaria in the USA, this highly venomous and colourful fish has undergone a spectacular explosion in numbers. In just a few years it has spread throughout the Caribbean, living sometimes in huge densities, feeding off Caribbean fish which seem to be naive to its stalking behaviour. Engorged and fatter than those living in balance with their surroundings in the Indo-Pacific, Caribbean lionfish multiply prodigiously, with unknown but much speculated effects. It will probably never be eradicated from the Caribbean.

Coral reef fishing

Fishing was described in terms of biomass reduction and biological effects earlier; this section looks at the effects when fishing is carried out on coral reefs and the difficulty of addressing the issue.

The vision of the tropical village living sustainably alongside its coral reefs with a man fishing for his family is an idyllic picture, but it is, sadly, largely a myth. There has probably never been a time, except in rare circumstances, of only a tiny human population in which fishing removed optimal numbers of fish in a sustainable way, leaving enough mature adults to allow for continued levels of fish in later generations. There are examples of traditional management which maintained, at least in the past, large fish stocks, but these were, for example, in royal preserves where the penalty for poaching was extremely severe, comparable to punishments for poaching in royal hunting parks around London centuries ago, where execution was the penalty. Today, where there are concentrations of people (and these concentrations are increasing rapidly) depletion of reef fish can be swift and extreme, disrupting the function of the reef, and thus its continuing ability to support human communities.

The problem escalates when increasingly desperate people try to catch the diminishing numbers of remaining fish, and in more and more destructive ways. Nets become snagged and lost, causing what is termed 'ghost fishing' for years or decades, and explosives may be used including those made from easily available fertilisers. Poisons including cyanide, and pesticides such as DDT are sometimes used. Commonly, large areas of reef are irreparably destroyed in the process and, whether illegal or not, profits can be large. It was estimated a decade ago that dynamite fishermen in one Asian country earned three times more than a local university professor, and many times more than the easily bribed local coastguard or policeman. When one area was wiped out, the poachers simply moved on.

We saw that unfished reefs might have an average of well over 7,000kg of fish per hectare while fish biomass on exploited reefs is commonly only one-tenth of that at best, sometimes only one-hundredth. Further, we saw that there is an almost complete absence of examples where fish exist in a biomass in the region of about 3,000 to 7,000 kilograms per hectare^{-1}, with most reefs today having biomass of about 1,000 kilograms per hectare^{-1} or less. Possibly such reefs have just never been found, or perhaps when a new reef system is discovered and is exploited it traverses that gap, the 'exploitation gap', very rapidly as the easiest components are caught and fish biomass plummets from the rich to the low state. There comes a point on many reefs where trying to fish becomes an increasingly fruitless endeavour.

Where the fish are removed from what is a very complex food web, a cascading collapse in other parts of the ecosystem can be seen, to the detriment of the whole reef. This has been demonstrated countless times, and we need think no further than the removal of herbivorous fish, so that algal growth is no longer suppressed. Add in some sewage to encourage more algal growth, and the result is usually a complete domination of algae at the expense of corals and habitat complexity.

The relentless downward spiral of the ecosystem might be inevitable, explained by analogy with the Ponzi effect. Charles Ponzi conducted a remarkable financial scam in the 1920s, whereby people invested in his scheme with the promise of high interest on their capital. Large returns were made, but the money used to pay them was not from interest earned but from the capital given by later investors. There comes a point when such schemes inevitably unravel. What has this to do with exploiting reef fish? There is no suggestion of any fraud, but the principle is as follows: the lowest level of fishing exploitation might be a man in a boat, fishing for his family. To make life easier (nobody can blame him) he purchases a bigger boat and engine. How does he pay for it? From the capital provided by the reef, namely from selling more fish; he has to catch more fish every day just to pay for it, before he has any money left over to eat. He forms a consortium in which people are employed to process this larger quantity of fish. He pays their wages by catching more fish (more reef capital). The village funds a small pier, then a larger harbour of the sort seen all around the tropical world, then perhaps a freezer plant. So, more and more fish are taken. Everybody knows that the biggest fish have the greatest monetary value; they literally provide the most capital, yet these are also the fish that also would have produced the greatest quantity of eggs (interest) by a great margin, which would produce the next generation of fish. Thus, the coastal society is not living off 'interest' provided by the reef, but is living off and depleting its capital until, as is commonly seen in much of the tropical world, very little remains.

The tragedy is that an economist will talk of 'investing' in fisheries—a freezer plant, say—but it is not an investment at all, it is a drawdown of reef capital. Interest could be possible, but only if fish can be sustained at high enough levels to support an equally abundant next generation, which does not appear to happen.

This chain of events is rooted in natural human behaviour. Nobody should blame local communities for wanting an easier or better lifestyle for themselves or their families—and this makes the problem all but intractable. It is Hardin's 'Tragedy of the Commons' for coral reefs.

A second key misunderstanding is use of the word 'sustainable'. Sloppy use of this grossly overused word is seen in any aid or development document today. Biomass in fished reefs is a small fraction of what it could be and, worse still, it is inevitably *sustained* in a depleted state for decades at a time. *Suppressed* is surely what is meant. Yet, so urgent and intractable is the problem, that phrases such as 'sustainable intensification' are now being introduced, which to any ecologist is an oxymoron.

It is not only coral reefs that suffer from overfishing. In all oceans of the world the major pelagic species that are hunted, such as tuna and its relatives, are declining. In the words of the eminent fisheries biologist Daniel Pauly, 'We still need to invent sustainability' with respect to fishing. The difficulty is that it is so easy to remove most fish biomass very quickly and simply. Several years ago it was calculated that the level of reef fishing occurring globally would require the reef equivalent of 2.5 more Great Barrier Reefs in order to maintain fishing at current levels. We don't have these extra reefs, of course, and indeed more reefs are degraded all the time.

There may be a way around the dilemma, and that is discussed in the final chapter. It should be remembered, however, that in any case, we cannot *manage* fish levels, as many planners seem to think. We can only manage the human behaviour affecting them.

Chapter 8
Climate change and reefs

Reefs are more affected by the damaging consequences of climate change than any other known ecosystem. This is not a problem for the future, as many think, but rather it is happening now. The damaging impacts of climate change add to, and synergistically multiply, the harmful effects already discussed. In this respect, coral reefs foretell the global condition perhaps more than other ecosystems.

Whereas diseases, pollution, and so on have had an enormous impact on the reefs of the world, all are in a sense fairly local, but together and singly they have now affected most of the world's reefs in such a way as to diminish their biodiversity and productivity. These 'local' impacts have been joined in the past couple of decades by truly global consequences arising from climate change. These exert effects in addition to those from the local factors, and they stem from rising carbon dioxide in the atmosphere, which leads to warming and acidification.

Warming and coral bleaching

Coral bleaching was the first manifestation of climate change affecting reefs, although initially many people did not really recognize it as such. Vast expanses of corals, commonly extending across the whole reef, simply turned white. First researched on the Pacific side of Panama in the 1980s, it was then observed to a massive extent throughout the Indian Ocean in 1998, in some parts of the Caribbean around this time too but mostly later in 2005, and in various parts of the Pacific and Australian region in a complex pattern since then. The phenomenon was reported in the 1870s but the main causes then were probably sewage and pollution. For a long time, climate effects stayed under most scientists' radar, and even scientific reviews commenting on reef decline written as late as the 1990s scarcely mentioned it. One influential review on potential vulnerability of coral reefs noted that remote oceanic atolls were amongst the most immune from degradation simply because of their low human populations and distances from large continental sources of runoff. However, massive and widespread ocean warming equally affected those distant reefs. Views had to change rapidly.

The first ocean-scale bleaching event occurred in 1998, when corals on reefs throughout almost the entire Indian Ocean turned white, following which corals and soft corals died. This extended to various depths depending on location, sometimes to as deep as 40 metres. We realized then that the main environmental drivers affecting reefs had shifted to something altogether different.

The term 'warming', however, is usually shorthand for several different variables. True, warming is most important, but alongside it is change in light intensity received by symbiotic algae. Changes to weather patterns cause changes to wind and sea conditions, such as those in the Indian Ocean warming event—for example, changes in wind patterns meant that the sea was glassily calm for long periods of time. As any underwater photographer knows, the brightness at any particular depth is different depending on whether it is stormy above or clear with flat calm seas, the latter letting through appreciably more light. The difference is sufficient to 'overload' the photosynthetic pathways in the corals' zooxanthellae.

Through thousands of years, corals in any area have become acclimatized to the ambient seawater temperature and light levels, and to the range of variation of each. Corals have adapted to certain conditions, and their optimum temperature happens to be only a couple of degrees cooler than temperatures that kill them. When temperatures remained significantly higher for weeks, and when the more intense light overloaded the photosynthetic mechanism, the symbiotic algae died. They were expelled, and then the corals died a few weeks later.

The concept of 'degree heating weeks' has proved very useful here. This is the number reached when we multiply the level of temperature rise over optimum conditions for corals by the number of weeks it remains high. Different species are, unsurprisingly, affected differently, but the number 10, resulting, for example, from five weeks at a temperature of 2°C above what is usual for that time of year, will generally prove fatal for most corals.

The coral may not die after bleaching. If the conditions are not too intense and do not last too long then re-acquisition of algal cells may take place and the coral may recover. But if it stays too warm for too long, re-acquisition of algal symbionts or re-growth of any that remained, does not happen; the coral colony cannot photosynthesize and will die. The visual bleaching arises because coral tissue itself is usually fairly transparent, and expulsion of zooxanthellae removes the pigments so that the white limestone beneath is seen. Dead coral tissue will slough off, again revealing pure white limestone. However, the bleached appearance does not last more than a couple of weeks because small filamentous algae, protozoa, and many other groups quickly colonize the newly available bare substrate, turning it dark again.

We know now that, in general, sea surface temperatures began to rise in the 1970s. But for a decade or two the change was small and could have been due to random fluctuations. After all, seasonal changes tend to be much greater than the underlying upward pattern; the 'noise' in the data is bigger than the 'signal'. By the end of the last millennium, however, the underlying trend of rising temperature, which by then was being vigorously researched, became clear. The trend continues: almost all of the warmest years of the last several thousand years have occurred in the last couple of decades. The warming trend has also accelerated to such an extent that it is predicted that bleaching could happen as frequently as every one or two years on most coral reefs in the next couple of decades. Given that a young coral needs on average to be about five years old before it can reproduce, bleaching events of this frequency will kill corals that are still too young to have reproduced.

Acidification

While warming has been overwhelmingly the most important cause of destruction of reefs during the last few decades, another factor that potentially could become equally important relates to the increasingly thick blanket of carbon dioxide we are weaving over the Earth. This causes acidification of the water. The pH of seawater is on the alkaline side of neutral: 8.3 is a common value. Carbon dioxide dissolves in water to become, first, carbonic acid, which is in equilibrium with the dissolved bicarbonate and carbonate. The buffering system works in a rather counter-intuitive way, but, in broad terms, a solution of carbon dioxide acidifies the water and reduces the amount of carbonate available to corals for limestone deposition. Some estimates already suggest that coral growth rate has decreased significantly because of this effect. The pH values themselves do not appear to be much changed, in that oceanic pH typically has dropped by about 0.1 units, but this is a logarithmic scale and equates to a 30 per cent increase in the number of hydrogen ions, which crudely means level of acid. (We should really talk of 'reduced alkalinity' rather than acidification, but that is perhaps too clumsy a phrase.)

The story is complicated because atmospheric carbon dioxide takes 20 to 40 years to reach equilibrium with the oceans—there is a time lag in other words, sometimes called a 'legacy', before carbon dioxide already in the atmosphere has its greatest effect. Taking this into account, it is thought that in 20 or 30 years the number of hydrogen ions in seawater will be three times higher. Our atmospheric blanket is not only increasing but its *rate* of increase is rising too. Pre-industrial levels of atmospheric carbon dioxide were around 280 parts per million (ppm). Unfortunately, it has been calculated that (given equilibrium with the oceans) a level of 350 ppm would be grossly inhibiting, if not lethal, to coral reef growth. In 2013, the level reached 400 ppm.

Carbon dioxide, then, causes both warming and acidification, both effects proven to have caused harm, and both predicted to continue on this damaging trajectory. One consequence of severely damaged or destroyed reefs is that they erode, because the bioeroding organisms living on the reefs are not also destroyed; indeed, their effects accelerate erosion simply because of the increased amount of substrate the bioeroders have to work on. Coral reefs provide the breakwater around thousands of islands and thousands of kilometres of coastline so, unsurprisingly, many governments are showing growing concern.

Sea levels and hurricanes

Shoreline erosion as a consequence of destruction of the reefs is often conflated with the observed rise in sea level. Sea level is rising at an average of a few millimetres per year, although this varies significantly depending on location. The averaged sea level rise in the 20th century due to warming was 17 centimetres and today this rate is accelerating. This is partly due to a greater amount of water in the oceans because of melting ice, and partly due to thermal expansion of the warmed seawater. The rise in sea level is causing serious problems for many islands, although the rise by itself may be less important than the progressive destruction of the coral reefs, which essentially provide offshore breakwaters. After all, a healthy reef can grow vertically, as indeed almost all of them have done extensively throughout their existence, but in order to do so the corals and their interlocking ecosystem must be healthy. Whatever the relative importance of sea level rise and reduction of offshore breakwaters, wave energy striking shorelines is increasing significantly, with important consequences to coastal infrastructure such as roads, facilities, and housing in many parts of the coral reef world.

Warming is having other effects on the weather too. Hurricane patterns appear to be changing. To develop into a hurricane, cyclonic winds need to be over ocean water of temperatures greater than about 26.5°C. In the Caribbean, where this has been researched extensively, indices called the Accumulated Cyclone Energy show significant rises. Hurricanes seem to be trending towards being larger, longer lasting, or stronger, rather than towards increasing in number, and this leads to more destructive energy. Most coral reef areas across the world have experienced hurricanes throughout their existence and have developed accordingly, but we have little idea how coral reefs will respond to increased levels of hurricane energy. Interestingly, the Atlantic coast of South America, which supports a uniquely distinctive collection of coral reefs, had never experienced hurricanes until 2004—when Brazil experienced its first.

Chapter 9
Doing something about it

Identifying and presenting the main issues effectively

It is helpful, when talking to decision-makers, especially civil servants, ministers, and the like, to use human analogies where possible. Most of these people, who have ultimate responsibility over the fate of coral reefs, have probably never even seen any reefs. So a useful analogy might be disease: we can survive many diseases if we are well fed and healthy, and mostly we will recover with no effects. But a person is far less likely to survive if afflicted by, for example, measles, malaria, cholera, and influenza all at once—*and* is malnourished too. Today this analogy can be applied to reefs. Not only are reefs affected by the ecological consequences of overfishing, pollution, sedimentation, and the rest, they are also stressed by higher than optimal ocean temperatures and the higher than optimal acidity levels of the water. This is why about a quarter of reefs are already dead, probably irrecoverably so as far as we can tell, and a further half are affected to various degrees of severity. This leaves only about a quarter that can be considered to be in a good state, with perhaps only a tiny proportion that can be considered to remotely resemble the state of reefs several decades ago.

The ecologically stable state of a coral reef is, quite naturally, presumed to be a condition in which corals dominate. But this is not necessarily the case. We know that there are alternative stable states, as illustrated in Figure 14. Imagine that a coral-dominated reef is represented by the black ball settled in the valley in the top sketch of the figure, separated from a lower valley by a hill. Stress the ball (the reef) a little with a push and (as shown by the arrows in the top ball) it will simply wobble in the valley and then, unless the shove was very hard, it will settle back into the same valley when the stress is removed. These little pushes represent the normal occurrences that reefs experience daily and annually. If we then imagine two ecological impacts acting on the reef (the lower of the two sketches of valleys), one perhaps being input of sewage, which stimulates seaweed—this might be represented by a reduction in the height of the valley—and the second impact might be overfishing, which might be represented by an extra hard shove to the ball. The ball will roll over into the next, lower valley, which in this case represents a limestone reef platform covered by seaweed. This state seems to be perfectly stable and the ball will not return to the upper valley (the coral dominated state) by any means that we know. Even removing the two stresses of sewage and overfishing will not help to do that. This in engineering terms is known as the hysteresis effect (the lagging of an effect behind its cause), and it appears to apply to reefs too (bottom sketch in Figure 14). Unfortunately, seaweed dominated limestone platforms are becoming increasingly common where stressors exist above certain fairly low levels. It is presumably possible to regain a coral dominated state somehow because, after all, coral reefs do still exist—it may just be that a much longer period of time is required without the stressors for this to happen.

Debate continues about whether seaweed domination occurs primarily as a result of herbivore removal, pollution, or raised temperature killing the corals. However, the answer undoubtedly depends on the specific location, and even sections of coast fairly close to one another can suffer from different stressors. Furthermore, the stressor that changes the ecosystem may be

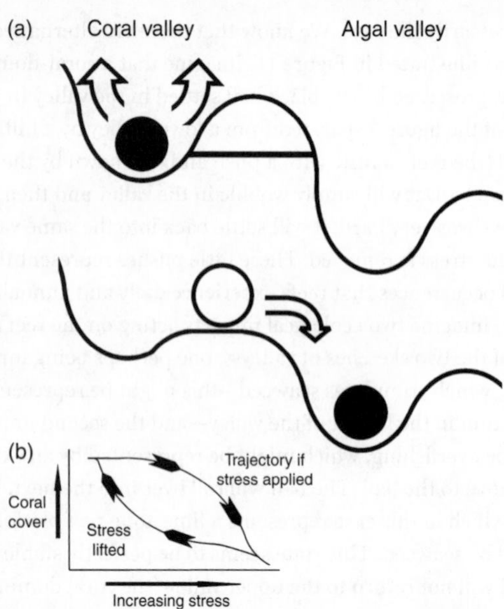

14. (a) Conceptual diagram of different ecological states of a reef. The valleys represent different ecological states: coral dominated on the left, algae dominated on the right. Top: the black ball is in the coral valley (a healthy reef); small shoves can be coped with: the ball returns to its stable valley. Below: when pollution reduces the height of the valley wall, and another stressor gives the ball a harder shove, it rolls into the lower valley—an algal dominated state. It does not have any known inclination to return, against gravity, to the coral dominated valley. (b) The hysteresis curve (an engineering term) is another way to represent shifts in community structure. The top curve shows stability as stress increases showing no change in coral cover until the system collapses. Removing the stressor, however, does not permit the same trajectory back to health, needing much more time before recovery is seen

different from the stressor that might later maintain it in the altered state for several decades.

Extremely degraded reefs are now not uncommon. I have dived for two hours on reefs made famous in the 1960s for being extremely rich in diversity, and not seen a single live coral or fish—this was shortly after the major warming in 1998. The rubble left was sedimented and covered with a cyanobacterial and algal film. It has been said that the switch to such forms from complex reef life is like moving backwards in evolutionary time, indeed even to Precambrian states, and is a 'slippery slope to slime'.

Despite the gloomy prognosis there may be glimmers of hope in several areas. Deeper zones on most reefs have escaped some of the worst damage and these can contain reservoirs of corals that could repopulate many shallow areas. Also the Western Pacific has not heated up as much as might have been expected, possibly because of a feedback mechanism operating in what is known as the Pacific Warm Pool, whereby water evaporates sufficiently when warmed, and the increased evaporation helps to cool the ocean, possibly also causing the formation of more clouds; and increased cloud cover has been linked to a reduction in warming. We must hope that this feedback effect will increase.

Adaptation by corals to warming and other factors is a much hoped for solution too. We know for example that many species of coral in natural warm basins such as the Arabian Gulf can survive annual high temperatures of 33–34°C, whereas the same species of corals in the Indian Ocean are killed at 29–30°C. The Arabian corals have clearly adapted—some of them at least, not all—as coral diversity is much lower there. The mechanism involved, and whether the adaptation is by the coral host or its algal symbionts, is being intensively researched. In a sense it doesn't matter—adaptation has clearly occurred there, so perhaps we can assume that it can happen more widely. However, such natural adaptation has had several thousand years to take place, and the rate of change seen today suggests that there simply might not be the time for this adaptation to occur before the reefs die out.

The processes involved in these examples are poorly known, and anyway they do rely rather too much on hope—hope that increased evaporation will form more clouds; hope that coral adaptation might occur in time. But hope is no way to manage a planet! There are things we can do to effect change, though none will be easy.

Solutions

We must first understand the 'shifting baseline syndrome'. This phrase was coined 20 years ago in relation to fishing by the fisheries biologist Daniel Pauly, and refers to how our expectation of what a healthy ecosystem should look like changes over time. I have suggested that it applies equally to the condition of coral reefs. Any 'baseline' measured by a practitioner today (and so-called baseline surveys are commonplace) will be markedly different from the baseline that might have been measured long ago. Indeed, there is commonly no easy way of determining what a studied reef's true baseline condition might once have been. Attempts have been made to try to determine this baseline in many areas, but practitioners have concluded that the term 'natural' has commonly been used to refer to conditions when a particular observer first studied the area, and this might be very different to what it was before then. Even a small stress on a reef can often tip it over the edge into a different phase state. That level of stress would not, perhaps, have had a noticeable effect 50 years earlier, but today the stress is acting on a reef already stressed to an undeterminable degree. That reef's 'elastic' has been stretched nearly to breaking point already, so to speak, without us having known it.

Two groups of solutions offer the most hope, in my opinion. At the local scale, we need to control pollution, fishing intensity, and so on—all easy to suggest but very difficult to achieve, particularly in countries which are poor, or have little understanding of the natural systems that support them. Therefore these changes require education and management. But we must understand that we cannot manage a coral reef, or any other marine system, come to that, and it is pure hubris to think that we can. All that we can do is to manage human impacts on the system and, where we can do that effectively, several aspects of reef health could start to improve. Furthermore, the *effects* of climate change would also be significantly delayed. The manager who thinks there is no point in, say, controlling fishing or pollution because climate change is going to damage the reef systems anyway needs to know that time can indeed be bought, perhaps a lot of time, if these preventative measures are taken.

The world needs, therefore, more management in reducing damaging activities: sewage and industrial pollution discharge, dredging, and fishing. These are effective measures, which are not hard to put in place in scientific terms. However, some of these damaging activities are supported by powerful groups who have vested interests and have the backing of well-funded lobbyists. Nonetheless, convincing arguments can be made that taking such measures, and allowing the reef to recover, will lead to the provision of more services to people—for example, the notable rise of food and fishermen's incomes in areas in the Philippines where this has been done effectively.

Properly conducted environmental assessments should be made before any major coastal changes and developments are carried out—most crucially, the construction industry needs to follow environmental advice. Then, a large project is likely to have minimal or temporary impact: a given country can then have its desired development and still retain a healthy reef ecosystem, which after all does provide both food and all-important protection from shoreline erosion. Sadly, many regions have ignored environmental assessments and, equally, many so-called environmental assessments have offered truly absurd solutions to local issues. Where development happens and the ecosystem is degraded severely, it is the local people who suffer, far more than the development group or the government. Social justice is not achieved if the environment suffers.

Environmental or habitat 'offset schemes' are becoming increasingly common. With these, a company planning an essential development incurring unavoidable negative impacts is required to restore some equivalent habitat that was damaged some time previously. There are of course problems in determining the principles on which the identification of an equivalent habitat, or an equivalent area, may be made, and these are somewhat arbitrary and difficult to agree. Such approaches may help reefs less than many other habitats, however, because one problem with reefs is that it has proved impossible so far to restore a damaged reef to anything like its original level. This has meant that any increased damage to reefs can only be offset by restoring quite different types of habitat, which is not good from the coral reef perspective.

The second group of measures involves developing protected areas *and* areas that are sacrificed to food production, as happens on land, instead of all areas being used for food in haphazard and decreasingly effective ways. The world needs more completely protected areas, and several agencies and individuals have estimated how much of the reef (or of any habitat) should be properly protected. By 'protected area' I mean one in which no extracting or destructive use is permitted, enabling proper refuges and seeding areas to develop. Estimates of the area needed range from 10 per cent to 40 per cent of the total, an alarmingly high proportion perhaps. Special pleading for exemptions, most commonly in favour of continued fishing, are common whenever a marine reserve is proposed but, when these exemptions have been granted, they have almost completely nullified the attempt to create a marine reserve. Marine reserves must be chosen and sited carefully, and then factors such as the state concerned actually having the ability to operate effectively in the first place must be taken into account—many do not.

An obvious limitation to protected areas is that, if one has a significant human population nearby, no amount of control can keep hungry people from impacting on it and nullifying any potential benefit. This means that sites must be carefully chosen. Where there are communities too large to permit any hope of conservation, then huge adjacent areas can be zoned for fish farming, for example—sacrificed, if you will, to the human need for food. People will obviously continue to try to obtain food from other areas too, so a zoning approach on a regional and ocean-wide scale will be needed if we are to prevent piecemeal decline throughout the oceans. The zoning approach has emerged only recently. It is based on pragmatism and runs counter to many traditional conservation approaches. However, since the latter is manifestly failing in most cases, the approach will likely gain credence in time. Where there are smaller communities, protected areas are more feasible and can work well. It is essential for local people to have a stake in the system; the principles and the point of conservation must be explained effectively and there must be a willingness amongst the communities to make it work. This has been achieved in an increasing number of places—as noted earlier, for example, in connection with villages in the Phillipines, where fish stocks and fishermen's incomes have increased as a result. When this happens, the villagers themselves have become the best ambassadors for the idea.

A half dozen enormous marine protected areas have been declared in the Pacific and Indian Oceans whose remoteness and other circumstances mean that they do not have huge direct human pressures and they also have governance that permits them to function. These large areas are likely to prove invaluable as they have an additional benefit, which is that the component reefs within each are interconnected in a genetic sense. Connectivity is important not only in a very large protected area but may also be critical in any coherent network of smaller protected areas too. Larvae can travel between reefs depending on the currents and seasons; different areas of reef might be mutually supportive.

The elephant in the room

Carefully avoided in many scientific discussions, conferences, government reports, and papers is the issue of human population. Indeed, in many conferences it is deemed to be a subject that is out of bounds. Rising numbers of people, and their desire for higher standards of living, put increasing demands on natural resources. More people are chasing a fixed or declining stock of reef resources: the area of the planet on which coral reefs can grow is limited, after all. In one sense it is really that simple. Some places have a human population doubling time of only 15 years, which reflects medical advances and its highly desirable accompaniments such as increased survival of people, especially infants. However, this means that current scientifically calculated solutions for a particular section of reef shoreline, for example, are negated when the population doubles. Thus the solution is no longer a scientific one, but has become largely a social and political one, and one of planning or zoning reefs and other resources as noted above. Human numbers are a part of the equation, and if we ignore any part of an equation then we cannot solve it. Several organizations have found that women's education in poor rural areas and their desire for family planning control is a crucial unmet need, which, when it is met, has positive effects on all aspects of the region—from improved human health and welfare, to improved capability of habitats, including reefs, to sustain the communities.

Degraded and overexploited reefs have a very high social cost. We can only make a rough estimate of how many people are malnourished or have died because their local reefs can no longer support them, but the number is already in the millions. It is difficult to understand why such massive effects are largely unreported by the general media and remain unknown to most. The issue of food security is paramount today, and it dominates many environmental documents. Estimates of mortality of children under 5 from malnutrition are 54 or 56 per cent; such children may at first glance have been killed by a particular disease, but these estimates suggest that it was malnutrition that made them vulnerable to the disease in the first place. The Food and Agriculture Organization of the UN reported: 'in general the countries that succeeded in reducing hunger were characterized by rapid growth in economic and agricultural sectors. They also exhibited slower population growth.'

Costs and values

It may be commonplace today to draw attention to the difference between the value of something and its price (Oscar Wilde's famous dictum), but the distinction is still overlooked. Average cover by coral on reefs has, for a couple of decades now, been declining at a rate of 1–2 per cent a year over most of the world. Simple maths shows quite clearly that the prognosis is terminal unless *values* are reassessed. As a scientist who has researched reefs for nearly 40 years, I do not want my work to be merely a postmortem of what reefs once were. Unfortunately, this is all too likely to be the case. There is insufficient blending of human activity with the needs of the habitats that we depend on. It is now a political and social problem to solve, I believe, not simply a scientific problem. Otherwise, we would have solved it: we know what the problems are in sufficient detail to have achieved that.

Never has it been more critical to arrest this global trend. At present rates of degradation, coral reefs will probably be the first major ecosystem to become ecologically extinct in this Anthropocene epoch—a quarter of them have already gone after all. Even though scientists instinctively show caution in their pronouncements, this possibility of near complete disappearance is being raised increasingly frequently, indicating how concerned we are with present trends. There is also, remember, the 'legacy time' to consider, the 20–40 years' delay described earlier before the full impact of factors such as water acidification become apparent. The decline of reefs is not a smooth trajectory in any area, of course; it commonly shows periods of stasis interspersed with steps of deterioration. There is the evidence of threshold effects too: only a tiny additional shove is needed to push the black ball of Figure 14 over the hill into the algal valley once it is already positioned near the top. This is well illustrated with effects of temperature on reefs—the ecologist Nancy Knowlton providing a vivid analogy: 'when the thermostat is turned up one notch people tend to expect one notch's worth of additional heat, not a house in flames'. The episodic notches upwards in temperature of the oceans and consequent steps downwards in coral cover on reefs follow this pattern.

It still may be possible to arrest the decline and so continue to receive benefits from the flourishing world of coral reefs, but the political sector is where the change must come from. We know the solutions in scientific terms, some are simple physics and many are uncomplicated biology, but we don't know what to do in a political sense. My views here are not new, and I ended a book I wrote 30 years ago with these words that are more urgently applicable today:

> Someone once said in reference to that other tropical wonderland, the rainforest, that its destruction, because of short term negligence and profit, was like stripping a Rembrandt for its canvas. With a coral reef too, its assets are priceless and depend on it being and remaining a mosaic of connected form and living pattern.

We must learn to apply what we know much faster than we have done in the past. We know enough to save coral reefs from terminal decline. We must lead politicians to change, at least as fast as our climate is changing and as fast as people are stressing this ecosystem, if we are to stay ahead of the game. Many of these stresses affect most other ocean ecosystems too—we should remember that life on land has never existed without life in the sea, though life existed in the oceans for many millions of years before there was any life on land. People have drawn attention to the fact that many changes today, such as ocean acidification, mimic conditions thought to have existed during past extinction events, though this time mankind is driving the causes. The teeming life of the beautiful and complex ecosystems that are coral reefs can be saved, given the will.

Further reading

General and broad ranging books covering:

Chapter 1: Geology or biology?
Chapter 2: Ancient reefs and islands
Chapter 3: The architects of a reef
Chapter 4: The resulting structure—a reef

Birkland C. 1997. *Life and Death of Coral Reefs*. Chapman Hall.

Côté, I.M. and Reynolds J.D. 2006. *Coral Reef Conservation*. Cambridge University Press.

Dubinski Z. (ed.) 1990. *Ecosytems of the World 25: Coral Reefs*. Elsevier Science Publishing.

Fabricius K.E. and Alderslade P. 2001. *Soft Corals and Sea Fans: A Comprehensive Guide to the Tropical Shallow Water Genera of the Central West Pacific, the Indian Ocean and the Red Sea*. Australian Institute of Marine Science.

Hopley, D., Smithers, S.G., and Parnell, K. 2012. *The Geomorphology of the Great Barrier Reef Development, Diversity and Change*. Cambridge University Press.

Hutchings P.A., Kingsford M., and Hoegh-Guldberg, O. 2008. *The Great Barrier Reef: Biology, Environment and Management*. CSIRO Publication.

McClanahan, T., Sheppard, C.R.C., and Obura, D. (eds.) 2000. *Reefs of the Western Indian Ocean: Ecology and Conservation*. Oxford University Press.

Sale, P. (ed.) 2002. *Coral Reef Fishes: Dynamics and Diversity in a Complex Ecosystem*. Academic Press.

Sheppard, C.R.C. 1983. *A Natural History of the Coral Reef.* Blandford Press.

Sheppard, C.R.C., Price, A.R.G., and Roberts, C.J. 1992. *Marine Ecology of the Arabian Area. Patterns and Processes in Extreme Tropical Environments*. Academic Press.

Sheppard, C.R.C., Davy, S., and Pilling, G. 2009. *Biology of Coral Reefs*. Oxford University Press. Reprinted 2010.

Spalding, M., Ravilious, C., and Green E.P. 2001. *Atlas of Coral Reefs.* University of California Press.

Veron, J.E.N. 1995. *Corals in Space and Time: The Biogeography and Evolution of the Scleractinia*. UNSW Press.

Veron J.E.N. 2000. *Corals of the World.* (3 vols.). Australian Institute of Marine Science.

Veron J.E.N. 2007. *A Reef in Time: The Great Barrier Reef from Beginning to End*. Harvard University Press.

Chapter 5: Microbial and planktonic engines of the reef

Dinsdale, E.A., Pantos, O., Smriga, S., Edwards, R.A. et al. 2008. Microbial ecology of four coral atolls in the Line Islands. *PLoS One* 3: e1584.

Douglas, A.E. 1994. *Symbiotic Interactions*. Oxford University Press.

Kirchman, D.L. 2008. *Microbial Ecology of the Oceans*. John Wiley & Sons.

Chapter 6: Reef fish and other major predators

Cinner, J.E., Sutton, S.G., and Bond, T.G. 2007. Socioeconomic thresholds that affect use of customary fisheries management tools. *Conservation Biology* 21: 1603–11.

Friedlander, A.M. and Demartini, E.E. 2002. Contrasts in density, size and biomass of reef fishes between the northwestern and the main Hawaiian islands: The effects of fishing down apex predators. *Marine Ecology Progress Series* 230: 253–64.

Mora, C., Chittarol, P.M, Sale, P.F, Kritzer, J.P. et al. 2003. Patterns and processes in reef fish diversity. *Nature* 421: 933–6.

Chapter 7: Pressures on reef ecosystems

Ateweberhan, M., Feary, D.A., Keshavmurthy, S., Chen, A. et al. 2013. Climate change impacts on coral reefs: Synergies with local effects, possibilities for acclimation, and management implications. *Marine Pollution Bulletin* 74: 526–39.

Bruno, J.F. and Selig, E.R. 2007. Regional decline of coral cover in the Indo-Pacific: timing, extent, and subregional comparisons. *PLoS ONE* 2(8): e711.

Carpenter, K. et al. 2008. One-third of reef-building corals face elevated extinction risk from climate change and local impacts. *Science* 321: 560–3.

Mora, C. 2008. A clear human footprint in the coral reefs of the Caribbean. *Proceedings of the Royal Society B* 275: 767–73.

Mumby, P.J., Hastings, A., and Edwards, H. 2007. Thresholds and the resilience of Caribbean coral reefs. *Nature* 450: 98–101.

Nicholas A., Graham, J., and McClanahan, T.R. 2013. The last call for marine wilderness? *BioScience* 63: 397–402.

Riegl, B., Sheppard, C.R.C., and Purkis, S.J. 2012. Human impact on atolls leads to coral loss and novel communities: A modeling study. *PLoS One*. June, 7(6): e36921, pp. 1–11.

Wilkinson, C.R. 2008. *Status of Coral Reefs of the World: 2008*. Global Coral Reef Monitoring Network and Reef and Rainforest Research Centre.

Chapter 8: Climate change and reefs

Ateweberhan, M., McClanahan, T.A., Graham, N.A.J., and Sheppard, C.R.C. 2011. Episodic heterogeneous decline and recovery of coral cover in the Indian Ocean. *Coral Reefs* 30: 739–52.

Baker, A.C., Glynn, P.W., and Riegl, B. 2008. Climate change and coral reef bleaching: An ecological assessment of long-term impacts, recovery trends and future outlook. *Estuarine Coastal Shelf Science* 80: 435–71.

Burke, L., Reytar, K., Spalding, M., and Perry, A. 2011. *Reefs at Risk Revisited*. World Research Institute.

Donner, S.D., Heron, S.F., and Skirving, W.J. 2009. Future scenarios: A review of modeling efforts to predict the future of coral reefs in an era of climate change. In M.J.H. van Oppen and J.M. Lough (eds.). *Coral Bleaching*. Ecological Studies 205, Springer Berlin Heidelberg.

Fabricius, K.E., Langdon, C., Uthicke, S., Humphrey, C., et al. 2011. Losers and winners in coral reefs acclimatized to elevated carbon dioxide concentrations. *Nature Climate Change* 1: 165–9.

Gardner, T.A., Côté, I.M., Gill, J.A., Grant, A., and Watkinson, A.R. 2003. Long-term regionwide declines in Caribbean corals. *Science* 301: 958–60.

Hoegh-Guldberg, O., Mumby, P.J., Hooten, A.J., Steneck, R.S., et al., 2007. Coral Reefs under rapid climate change and ocean acidification. *Science* 318: 1737–42.

Hoegh-Guldberg, O. and Bruno, J.F. 2010. The impact of climate change on the world's marine ecosystems. *Science* 328: 1523–8.

Knowlton, N. and Jackson, J.B.C. 2008. Shifting baselines, local impacts, and global change on coral reefs. *PLoS Biology* 6(2): e54.

Riegl, B.M., Purkis, S.J., Al-Cibahy, A.S., Abdel-Moati, M.A., and Hoegh-Guldberg, O. 2011. Present limits to heat-adaptability in corals and population-level responses to climate extremes. *PLoS ONE* 6(9): e24802. doi:10.1371/journal.pone.0024802.

Sheppard, C.R.C. 2006. Longer term impacts of climate change. In I. Cote and J. Reynolds (eds.). *Coral Reef Conservation*. Cambridge University Press.

Veron, J.E., Hoegh-Guldberg, O., Lenton, T.M., Lough, J.M., et al. 2009. The coral reef crisis: The critical importance of <350 ppm CO2. *Marine Pollution Bulletin* 58: 1428–36.

Chapter 9: Doing something about it

Bradbury, R.H. and Seymour, R.M. 2009. Coral reef science and the new commons. *Coral Reefs* 28: 831–7.

Graham, N.A.J., McClanahan, T.R., MacNeil, M.A., and Wilson, S.K. 2008. Climate warming, marine protected areas and the ocean-scale integrity of coral reef ecosystems. *PLoS One* 3(8): e3039.

Hughes T.P., Graham, N.A.J., Jackson, J.B.C., Mumby, P.J., and Steneck, R.S. 2010. Rising to the challenge of sustaining coral reef resilience. *Trends Ecology and Evolution* 1282(25): 633–42.

Jackson, J.B.C., Kirby, M.X., Berger, W.H., Bjorndal, et al. 2001. Historical overfishing and the recent collapse of coastal ecosystems. *Science* 293: 629–37.

McClanahan, T.R., and Omukoto, J.O. 2011. Comparison of modern and historical fish catches (AD 750–1400) to inform goals for marine protected areas and sustainable fisheries. *Conservation Biology* 25: 945–55.

McClanahan, T.R., Graham, N.A.J., Calnan, J.M., and MacNeil, M.A. 2007. Towards pristine biomass: Reef fish recovery in coral reef marine protected areas in Kenya. *Ecological Applications* 17: 1055–67.

Mora, C., Andrèfouët, S., Costello, M.J., Kranenburg, C. et al., 2006. Coral reefs and the global network of marine protected areas. *Science* 312: 1750–1.

Riegl, B., and Purkis, S.J. 2009. Model of coral population response to accelerated bleaching and mass mortality in a changing climate. *Ecological Modelling* 220: 192–208.

Roberts, C.M., McClean, C.J., Veron, J.E., Hawkins, J.P., et al. 2002. Marine biodiversity hotspots and conservation priorities for tropical reefs. *Science* 295: 1280–4.

Russ, G.R., and Alcala, A.C., 2004. Marine reserves: Long-term protection is required for full recovery of predatory fish populations. *Oecologia* 138: 622–7.

Wilkinson, C. and Bernard Salvat, B. 2012. Coastal resource degradation in the tropics: Does the tragedy of the commons apply for coral reefs, mangrove forests and seagrass beds? *Marine Pollution Bulletin* 64: 1096–105.

Index

C

D

E

F

G

H

P

Q

R

S

T

U

V

W

Y

Z

GLOBAL WARMING

A Very Short Introduction

Mark Maslin

Global warming is arguably the most critical and controversial issue facing the world in the twenty-first century. This *Very Short Introduction* provides a concise and accessible explanation of the key topics in the debate: looking at the predicted impact of climate change, exploring the political controversies of recent years, and explaining the proposed solutions. Fully updated for 2008, Mark Maslin's compelling account brings the reader right up to date, describing recent developments from US policy to the UK Climate Change Bill, and where we now stand with the Kyoto Protocol. He also includes a chapter on local solutions, reflecting the now widely held view that, to mitigate any impending disaster, governments as well as individuals must to act together.

www.oup.com/vsi